W9-BXS-765

Senior Editor Miezan van Zyl
US Editor Karyn Gerhard
Senior Designer Mik Gates
Project Art Editor Jessica Tapolcai
Editor Michael Clark
Managing Editor Angeles Gavira
Managing Art Editor Michael Duffy
Production Editor Gillian Reid
Production Controller Laura Andrews
Jacket Design Development Manager
Sophia M.T.T.
Jacket Designer Akiko Kato
Associate Publishing Director Liz Wheeler
Art Director Karen Self
Publishing Director Jonathan Metcalf

First American Edition, 2021
Published in the United States by DK Publishing
1450 Broadway, Suite 801, New York, NY 10018

Copyright © 2021 Dorling Kindersley Limited
DK, a Division of Penguin Random House LLC
21 22 23 24 25 10 9 8 7 6 5 4 3 2 1
001–325074–Nov/2021

All rights reserved.
Without limiting the rights under the copyright reserved above, no part of this publication may be reproduced, stored in or introduced into a retrieval system, or transmitted, in any form, or by any means (electronic, mechanical, photocopying, recording, or otherwise), without the prior written permission of the copyright owner.
Published in Great Britain by
Dorling Kindersley Limited

A catalog record for this book
is available from the Library of Congress.
ISBN 978-0-7440-4445-4

DK books are available at special discounts when purchased in bulk for sales promotions, premiums, fund-raising, or educational use. For details, contact: DK Publishing Special Markets,
1450 Broadway, Suite 801, New York, NY 10018
SpecialSales@dk.com

Printed and bound in UAE

For the curious
www.dk.com

This book was made with Forest Stewardship Council™ certified paper—one small step in DK's commitment to a sustainable future.
For more information go to
www.dk.com/our-green-pledge

CONSULTANT

Professor Frans Berkhout is the Executive Dean of the Faculty of Social Science & Public Policy and Professor of Environment, Society, and Climate at King's College London.

CONTRIBUTORS

Clive Gifford is a Royal Society award-winning writer and journalist who has written and contributed to many popular-science and technology books.

Daniel Hooke studied climate change at University College London, with a particular interest in climate models and climates of the past. He has contributed to a range of climate change books for children and adults.

Adam Levy is a science journalist and climate YouTuber who holds a PhD in atmospheric physics from the University of Oxford.

CONTENTS

WHAT IS CLIMATE CHANGE?

WHAT IS CARBON?

POPULATION

FOOD AND RESOURCES

RISING CONSUMPTION

EFFECTS ON THE ATMOSPHERE

EFFECTS ON **LAND**

EFFECTS ON THE **OCEANS**

HUMAN COST

LARGE-SCALE SOLUTIONS

CHANGE ON A PERSONAL SCALE

CLIMATE CHANGE

Earth's climate has changed many times during its 4.54 billion-year history. Differing factors triggered these past changes, from alterations in the sun's intensity and variations in Earth's orbital path to volcanic activity and meteorite impact. Most of these climate changes took tens of thousands or millions of years to occur and some had a profound impact on the planet.

The current climate change we are experiencing is different from those that went before. After decades of denial and skepticism, the mounting body of evidence gathered by science is overwhelming. The planet is both warming at an unparalleled rate, and human activity, rather than natural phenomena, is the primary cause.

Within the past two centuries, industrialization, unprecedented population and economic growth, urbanization, deforestation, and pollution have wrought extraordinary changes on the land, oceans, and atmosphere of our planet. The principal driver of current climate change is the increased emission of greenhouse gases responsible for producing an accelerated greenhouse effect within the atmosphere. Climate change's consequences are multiple, complex, and diverse, with an array of repercussions felt to varying degrees across different parts of the planet.

Understanding the scale and scope of present and future impacts is part of the investigation into the interconnected nature of human activity with the planet and its resources and processes. This vital endeavor is highlighting just how much climate change will touch all aspects of human life and society, and how solutions, mitigating actions, and adaptation strategies are needed to manage and live with a changing climate, now and in the future.

W H A T
C L I M
C H A N

I S

A T E

G E ?

All environments on Earth, from the frozen poles and deep oceans to the scorched deserts, are connected by the climate. As scientists have modeled, measured, and recorded more of Earth's climate, using data from satellites and drifting ocean buoys, a clearer picture of these connections has emerged. In addition, by using data from the past to study natural climate shifts, such as ice ages, the influence of the global greenhouse effect has been confirmed. Current changes to the climate will ultimately determine the conditions for all regions across the world that people call home.

Whatever the weather
Weather conditions often vary within minutes, hours, or days, often making them more difficult to predict than climate.

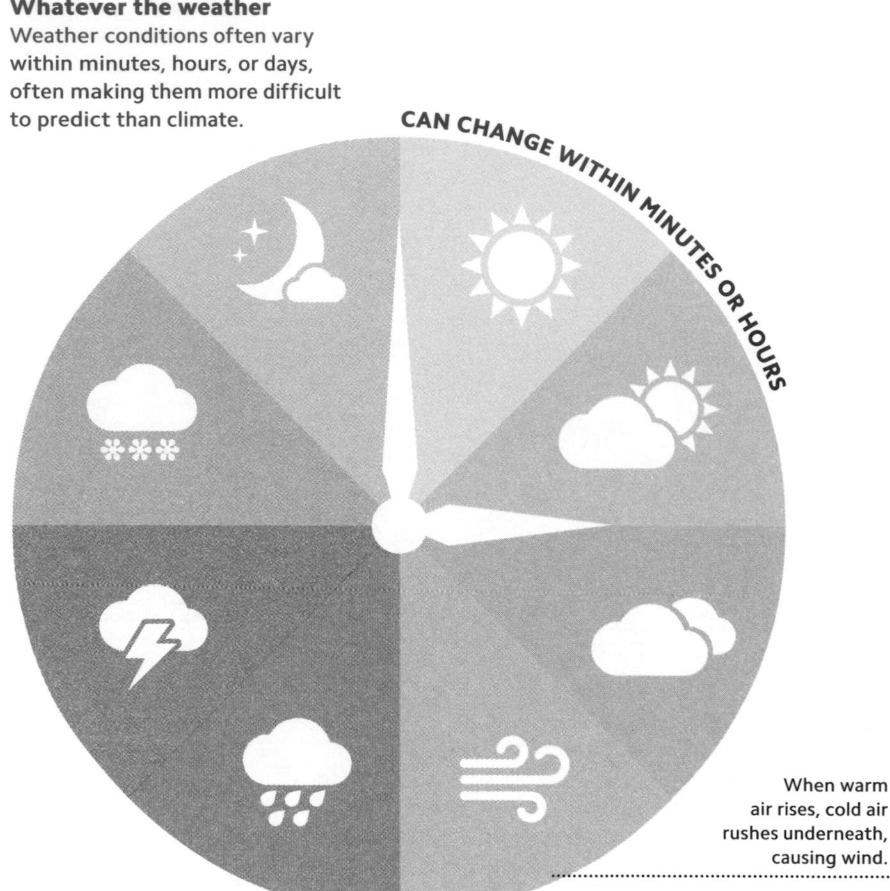

When warm air rises, cold air rushes underneath, causing wind.

WEATHER V

Weather is the short-term state of the atmosphere in a given location. It is caused by interactions between winds and water vapor. Across the world, everyone experiences the effects of weather, whether it is hot or cold, wet or dry, windy or calm. Climate is the average pattern of weather over a long period of time. Meteorologists typically define climate using a 30-year window.

Long-term shifts
Climates change slowly, over several decades. Human influence has sped up this process considerably.

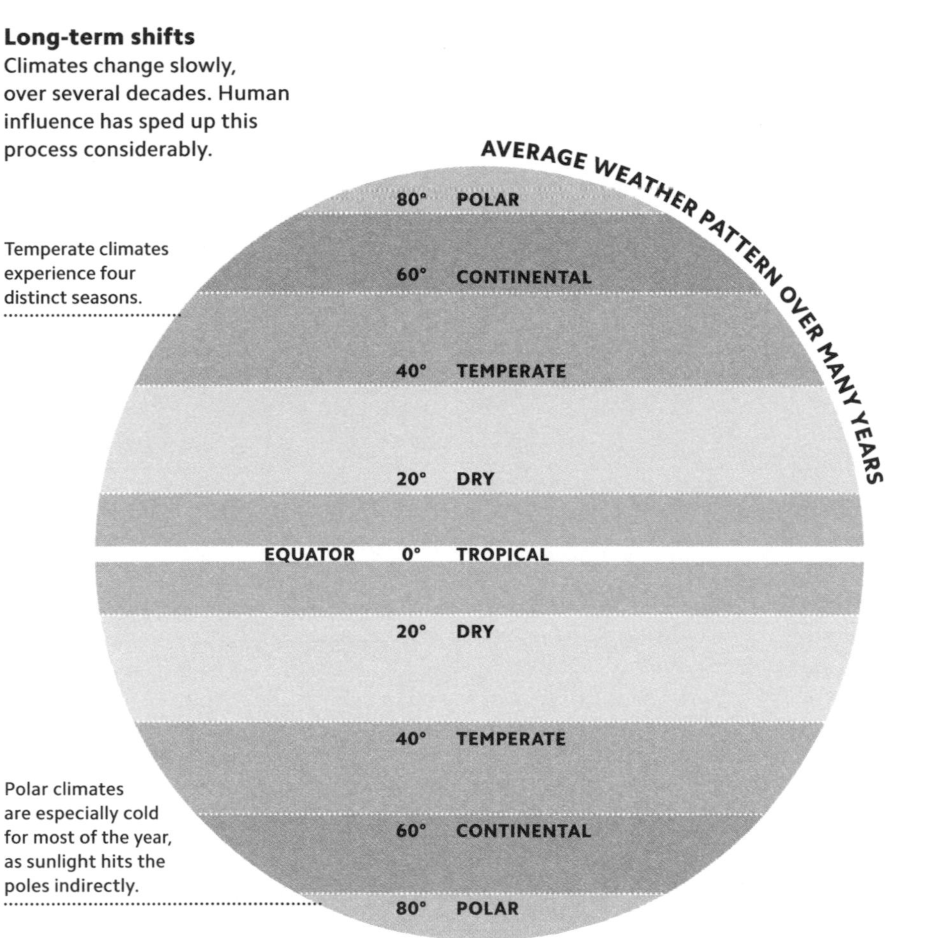

Temperate climates experience four distinct seasons.

Polar climates are especially cold for most of the year, as sunlight hits the poles indirectly.

S. CLIMATE

"Climate is what we expect, weather is what we get."
Robert A. Heinlein

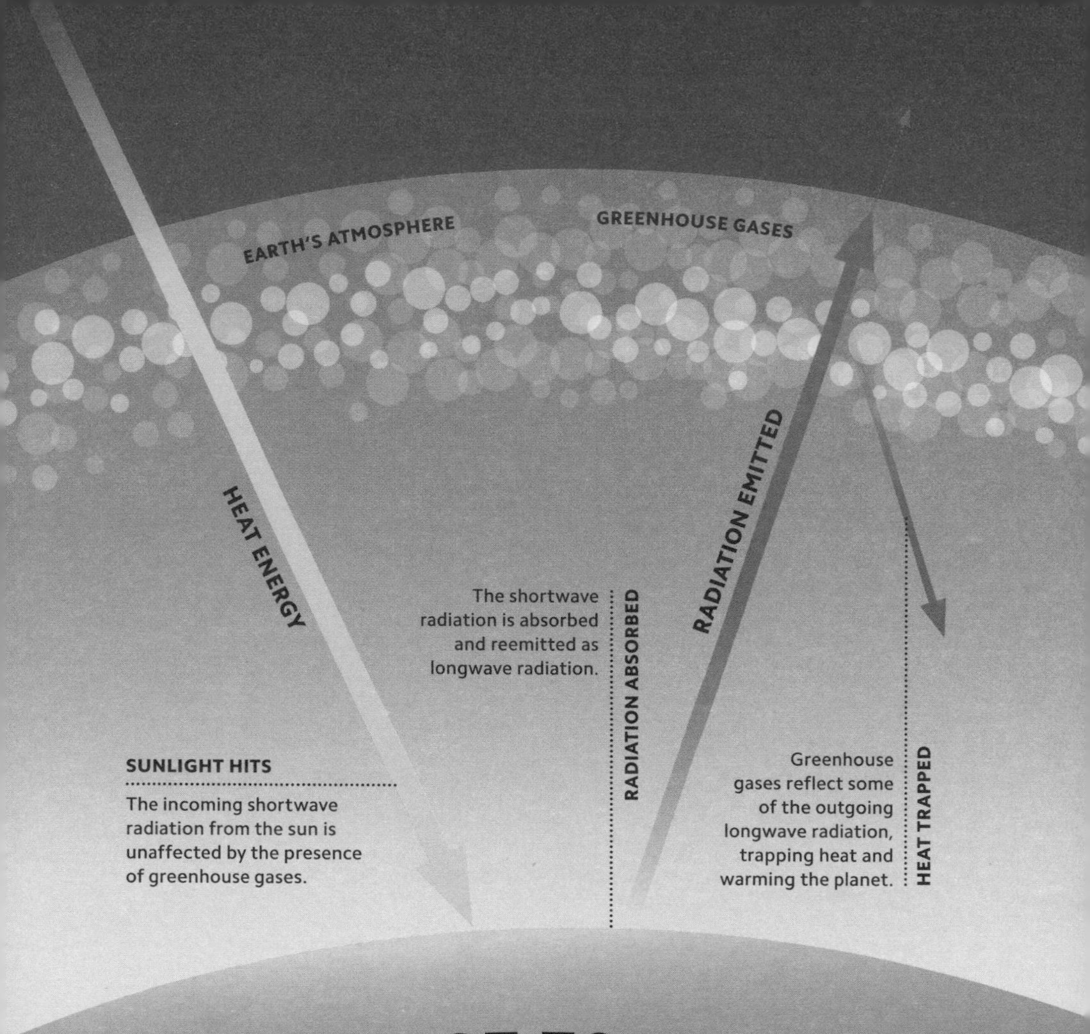

EARTH'S ATMOSPHERE

GREENHOUSE GASES

HEAT ENERGY

RADIATION EMITTED

The shortwave radiation is absorbed and reemitted as longwave radiation.

RADIATION ABSORBED

SUNLIGHT HITS

The incoming shortwave radiation from the sun is unaffected by the presence of greenhouse gases.

Greenhouse gases reflect some of the outgoing longwave radiation, trapping heat and warming the planet.

HEAT TRAPPED

TOO HOT TO HANDLE

The greenhouse effect is caused when energy from the sun travels through our atmosphere and is absorbed by the Earth, before being emitted outward as thermal energy. This energy interacts with greenhouse gases in the atmosphere, which reflect some of the energy back toward Earth, heating the planet. Greenhouse gases, such as those produced from burning fossil fuels, feed this process, trapping more energy and further heating the planet.

SOMETHING IN THE AIR

Of all greenhouse gases, carbon dioxide, which has been emitted on a vast scale, is the most damaging, as it typically stays in the atmosphere for 100 years. While methane and nitrous oxide are more potent, much less of these gases is emitted. Water vapor, the most abundant gas, is mostly not caused by humans. Ozone, a rare gas, contributes the least to the greenhouse effect.

OZONE
O_3

NITROUS OXIDE
N_2O

WATER VAPOR
H_2O

CARBON DIOXIDE
CO_2

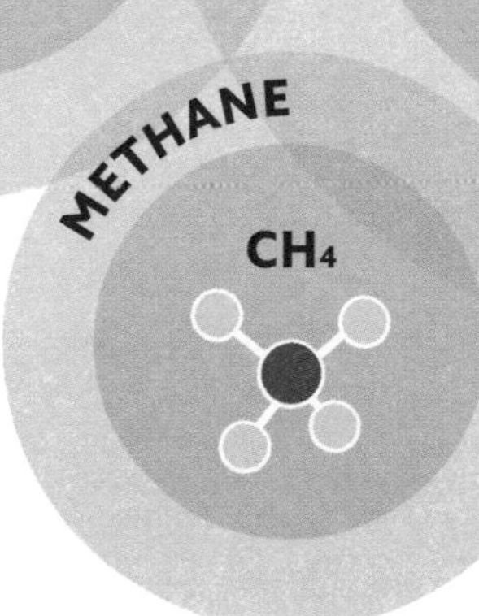

The main offenders
All of these gases are integral to the greenhouse effect. Atmospheric concentrations of carbon dioxide, however, dwarf those of methane, nitrous oxide, and ozone.

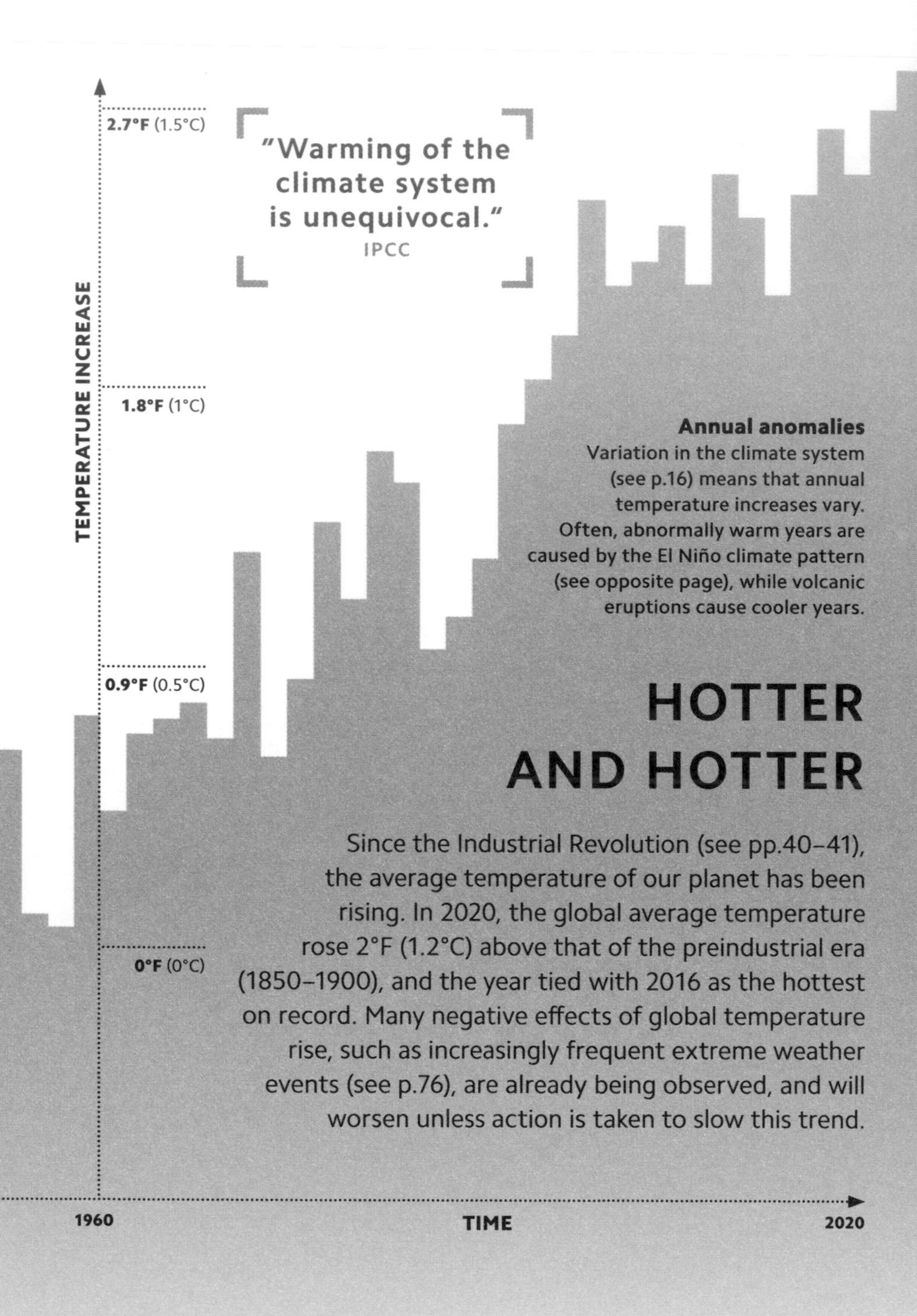

"Warming of the climate system is unequivocal."
IPCC

Annual anomalies
Variation in the climate system (see p.16) means that annual temperature increases vary. Often, abnormally warm years are caused by the El Niño climate pattern (see opposite page), while volcanic eruptions cause cooler years.

HOTTER AND HOTTER

Since the Industrial Revolution (see pp.40–41), the average temperature of our planet has been rising. In 2020, the global average temperature rose 2°F (1.2°C) above that of the preindustrial era (1850–1900), and the year tied with 2016 as the hottest on record. Many negative effects of global temperature rise, such as increasingly frequent extreme weather events (see p.76), are already being observed, and will worsen unless action is taken to slow this trend.

How El Niño occurs
Normally, strong easterly winds push warm surface waters to the west Pacific and cold waters rise in the east Pacific. During El Niño (see below) these winds slow, causing a buildup of warm waters across the Pacific.

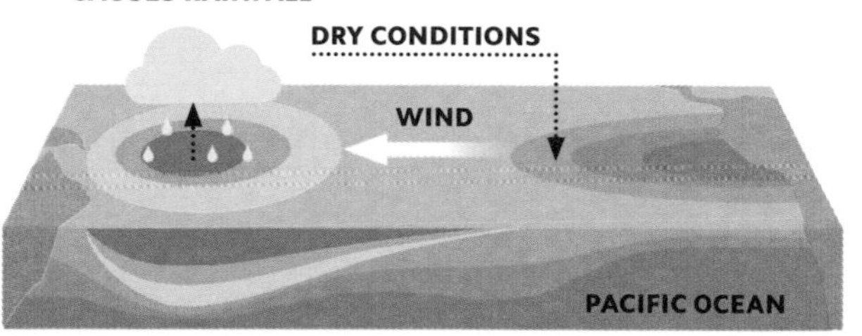

NORMAL CONDITIONS

DISTORTED CLIMATES

In the Pacific, a recurring climate pattern called the El Niño Southern Oscillation (ENSO) regularly influences weather around the world. In an El Niño phase, conditions for heavy rainfall spread across the Pacific, and a reversal in wind direction increases the risk of drought in areas such as India and Australia. ENSO is a natural cycle, but its frequency is projected to increase from once every 20 years to as often as once every 10 years if average global temperatures continue to rise.

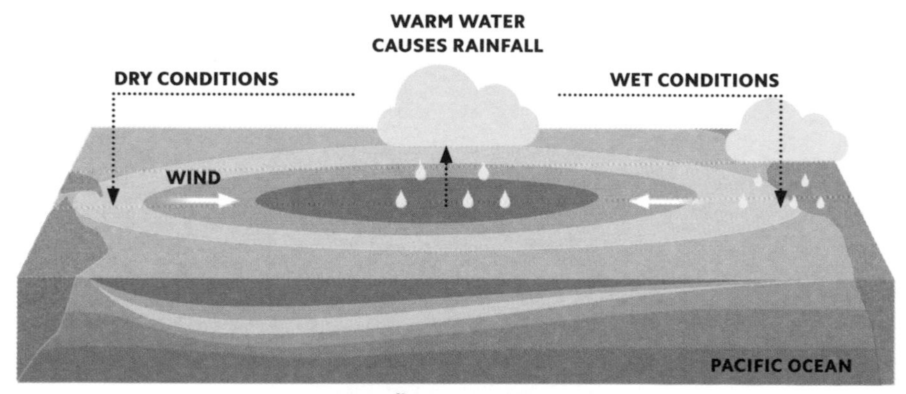

EL NIÑO CONDITIONS

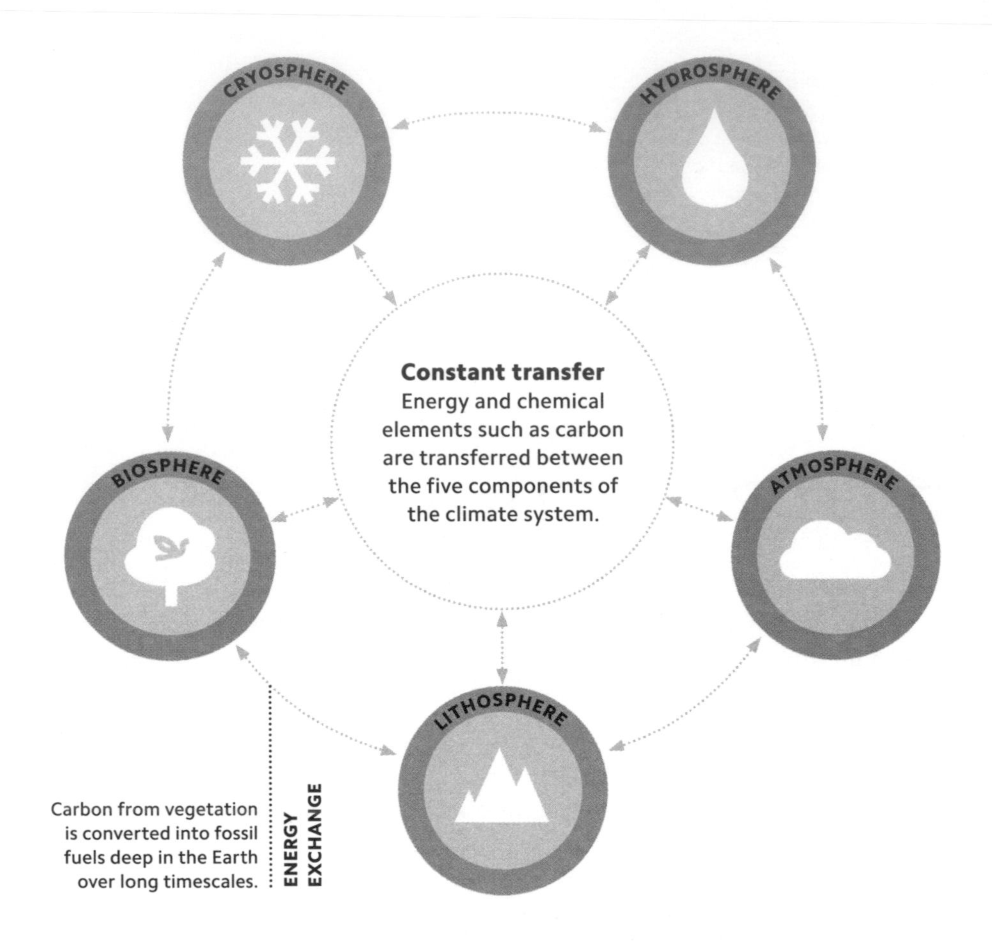

CRYOSPHERE
HYDROSPHERE
Constant transfer
Energy and chemical elements such as carbon are transferred between the five components of the climate system.
BIOSPHERE
ATMOSPHERE
LITHOSPHERE
Carbon from vegetation is converted into fossil fuels deep in the Earth over long timescales.
ENERGY EXCHANGE

A DELICATE BALANCE

The climate system is the relationship between Earth's five major components, all of which influence one another, and determine weather and climate patterns. Changes in the atmosphere (air) are often the most rapid, in comparison to the lithosphere (Earth's crust), the biosphere (living matter), the cryosphere (snow and ice), and the hydrosphere (water). Each component of the climate system is measured and analyzed.

OUT OF THIN AIR

Earth's atmosphere is 6,200 miles (10,000 km) thick and composed of five layers. In the troposphere, the closest 10 miles (16 km) where all weather occurs, greenhouse gases trap heat that warms the planet (see p.12). In the stratosphere, a band of ozone blocks harmful ultraviolet radiation from reaching Earth's surface. The mesosphere also protects Earth's surface, this time from meteors, which burn up on entry. The thermosphere, which is fifty times deeper than the troposphere, is the atmosphere's hottest layer, reaching up to 3,630°F (2,000°C). Above it, over thousands of miles, the exosphere merges with space, extending halfway to the moon.

EXOSPHERE
THERMOSPHERE
MESOSPHERE
STRATOSPHERE
TROPOSPHERE

Multiple layers
The atmospheric layers, in particular the troposphere, are thickest at the equator and become much thinner at the North and South poles. The five layers are determined by temperature.

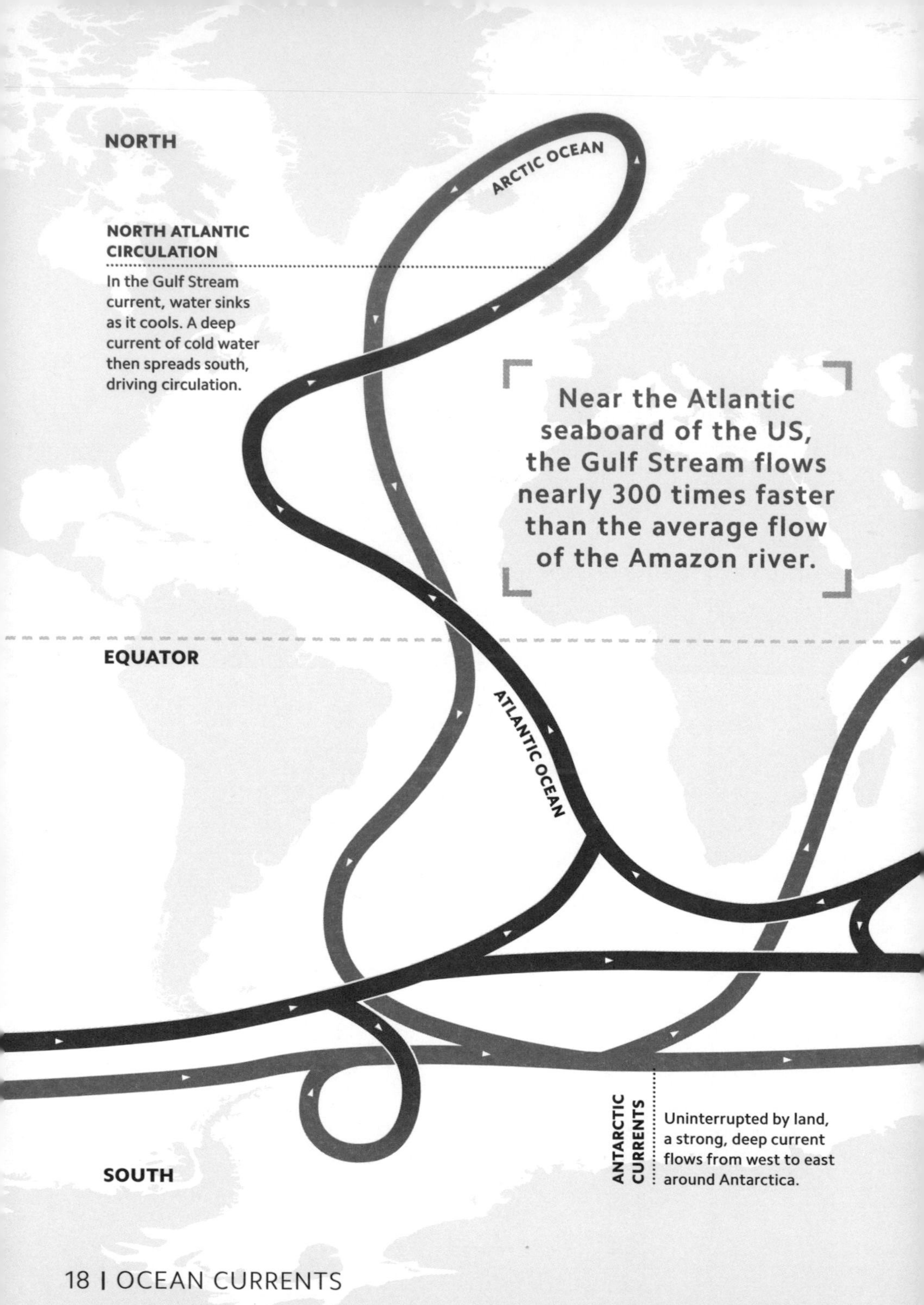

NORTH

ARCTIC OCEAN

NORTH ATLANTIC CIRCULATION

In the Gulf Stream current, water sinks as it cools. A deep current of cold water then spreads south, driving circulation.

Near the Atlantic seaboard of the US, the Gulf Stream flows nearly 300 times faster than the average flow of the Amazon river.

EQUATOR

ATLANTIC OCEAN

ANTARCTIC CURRENTS

Uninterrupted by land, a strong, deep current flows from west to east around Antarctica.

SOUTH

GO WITH THE FLOW

Ocean water constantly circulates in a global conveyor belt. This system drives heat around the world. Warmer surface currents (red) are primarily powered by winds, while cold deep currents (blue) are driven by variation in water temperature and saltiness. Some scientists think that rising global temperatures may cause the conveyor belt to slow down. This could worsen extremes in weather and temperature across the world.

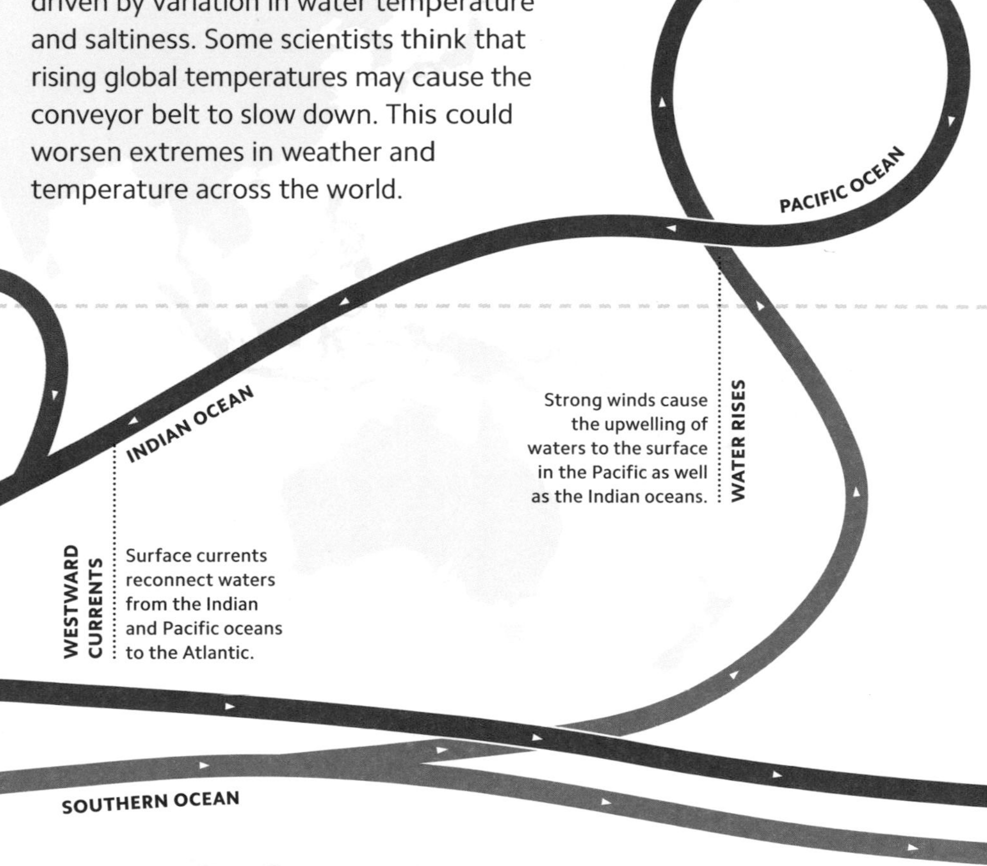

PACIFIC OCEAN
INDIAN OCEAN
WATER RISES
Strong winds cause the upwelling of waters to the surface in the Pacific as well as the Indian oceans.
WESTWARD CURRENTS
Surface currents reconnect waters from the Indian and Pacific oceans to the Atlantic.
SOUTHERN OCEAN

SPINNING OUT OF CONTROL

A tipping point is a threshold at which a fundamental shift occurs, making a return to the previous system impossible. Many parts of the climate system are feared to be in danger of crossing tipping points. For example, rising ocean temperatures have caused coral reef die-off to increase for many years. Experts warn that the regularity of coral bleaching could soon pass a tipping point, becoming near-annual. For this, and other tipping points, there is still time to act before it is too late.

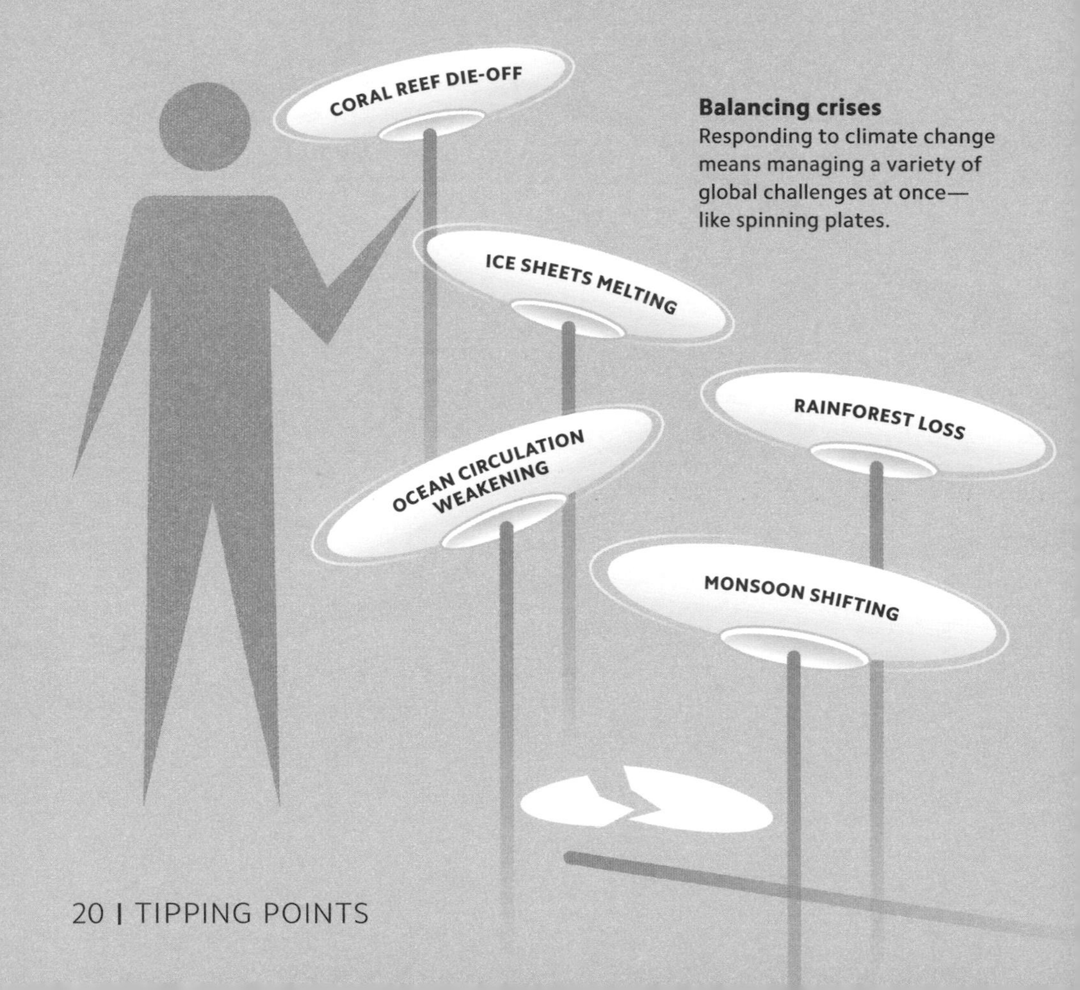

Balancing crises
Responding to climate change means managing a variety of global challenges at once—like spinning plates.

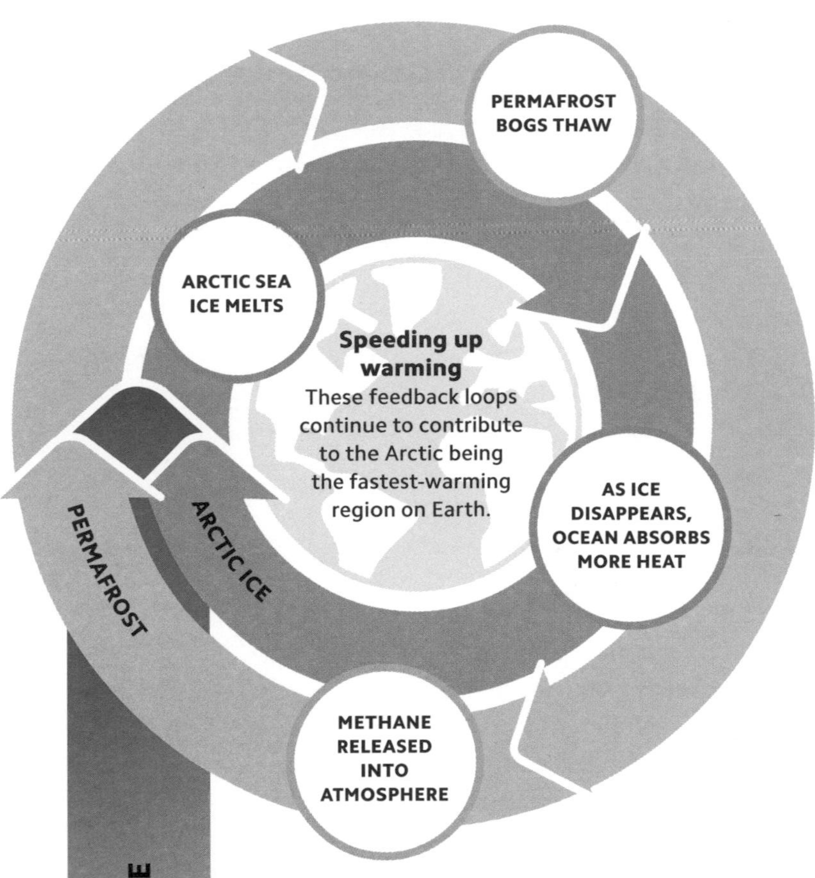

VICIOUS CYCLES

Feedback loops are effects of climate change that add to (positive feedback) or subtract from (negative feedback) the impact of the original change. While many aspects of climate change respond well to individual action, most feedback loops can only be mitigated by slowing the rise of average global temperatures. Without this, feedback loops, such as disappearing sea ice no longer reflecting sunlight, will continue to worsen.

Seeing the whole picture

Climate data is gathered from a variety of sources. Combining ground-based temperature measurements with those made from satellites increases confidence in the data.

GATHERING INTEL

Measuring the climate is key to understanding climate change patterns. Surface-level variables such as temperature, pressure, and rainfall are directly measured by meteorological stations. Traditionally, ocean data has been recorded by ships at surface level, but lately, purpose-built buoys have collected data from greater depths. Direct measurements higher in the atmosphere are made by floating balloons. Satellites have vastly increased the total area from which climate data can be collected, extending to the polar regions.

In 2016, an Antarctic ice core yielded ice that was 2.7 million years old—the oldest ever recorded.

Atmospheric record
Ice cores record any event large enough to leave traces in our atmosphere, such as nuclear explosions and volcanic eruptions.

WINDOWS INTO THE PAST

In search of a long-term record of CO_2 emissions on Earth, scientists in the 1950s began to drill deep ice cores in Antarctica to measure the concentration of CO_2 in thousands of tiny bubbles held in the ice. The longest continuous record yielded from an ice core, at 800,000 years, shows atmospheric CO_2 varied from 180 to 300 ppm (parts per million). Since the Industrial Revolution (see pp.40–41), however, this has increased rapidly, reaching 414 ppm in 2020. This rise is the result of human influence on the climate.

CORE CAPTURE

As snow compacts to ice, it traps bubbles of gas, which can remain frozen for millions of years. The deepest ice cores reach 1.9 miles (3 km) in depth.

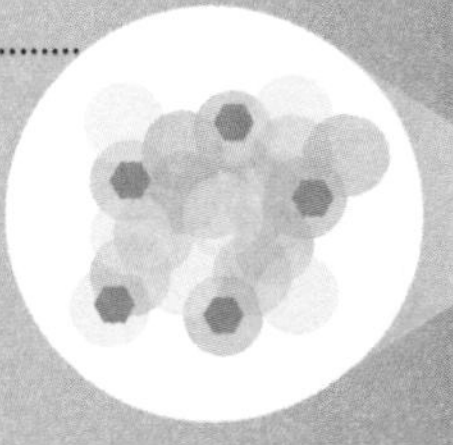

ICE CORE

USE OF NUCLEAR WEAPONS 1945–59

Unique radioactive isotopes derived from the 1945 nuclear bombing of Nagasaki, and subsequent nuclear weapons testing in the 1950s, have been found in ice cores.

KRAKATOA VOLCANO ERUPTION 1883

Significant volcanic eruptions deposit thin layers of ash across Antarctica, which are prominent enough to feature in ice cores.

MOST RECENT ICE AGE c.18,000 BCE

Ice core CO_2 levels reach minimal values during past ice ages, indicating a strong relationship between the prevalence of atmospheric CO_2 and temperature.

How climate models work
Using supercomputers that make trillions of calculations per second, climate models measure the horizontal and vertical transfer of energy, moisture, and carbon between their grid cells.

> A 2019 study of 17 global temperature projections made since 1970 found 14 to be exactly accurate.

RUNNING SIMULATIONS

Climate models divide Earth's land, atmosphere, and bodies of water into a three-dimensional grid of cells. Motions within the climate system can be vertical, such as the rising of water, or horizontal, such as wind movement. As computing power has increased, climate models have become more advanced, incorporating more complex processes such as ice-sheet dynamics. By inputting differing levels of human CO_2 emissions, scientists can simulate the global climate for past, present, and future conditions. Despite relying on complex uncertainties, climate models remain a useful tool to help us understand climate change.

FROM THE SURFACE

Climate models calculate the exchange of moisture, temperature, and materials such as dust and carbon between the ground and the atmosphere.

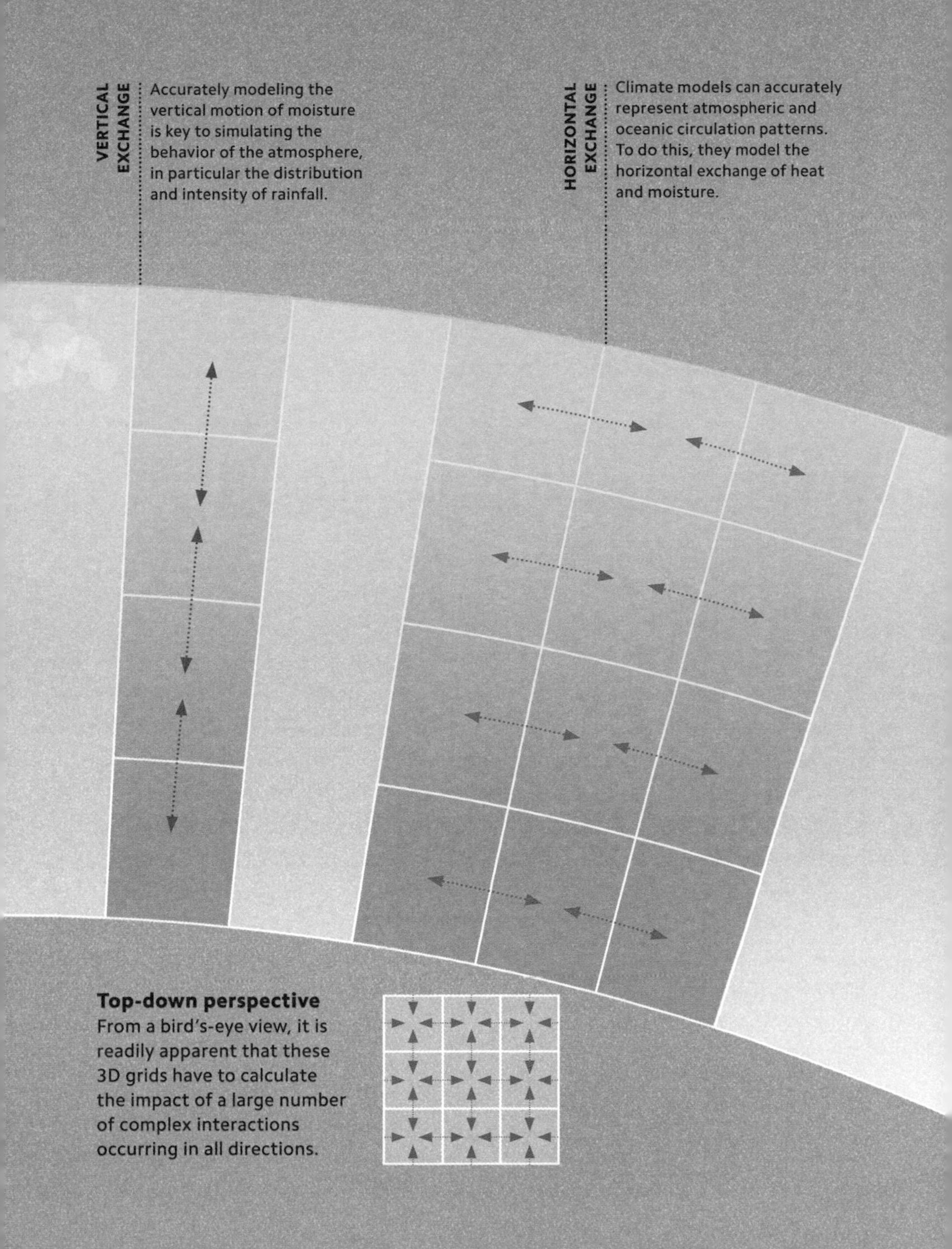

VERTICAL EXCHANGE
Accurately modeling the vertical motion of moisture is key to simulating the behavior of the atmosphere, in particular the distribution and intensity of rainfall.
HORIZONTAL EXCHANGE
Climate models can accurately represent atmospheric and oceanic circulation patterns. To do this, they model the horizontal exchange of heat and moisture.

Top-down perspective
From a bird's-eye view, it is readily apparent that these 3D grids have to calculate the impact of a large number of complex interactions occurring in all directions.

WHAT CARB

I S

O N ?

Carbon is an element that moves naturally between stores, principally the atmosphere, oceans, vegetation, and solid Earth. Carbon stored in the atmosphere, in the form of greenhouse gases, determines the strength of the greenhouse effect. Human processes have accelerated the transfer of carbon from stores deep below the Earth's surface, as fossil fuels, to other stores in the atmosphere and oceans. To limit the average global temperature increase in line with the Paris Agreement, atmospheric concentration of greenhouse gases must stabilize, and the human activities producing these gases need to be decarbonized.

FUEL TO THE FIRE

Burning fossil fuels is the biggest contributor to human-caused greenhouse-gas emissions. Coal and gas are mainly used in power plants to produce electricity, with a small portion used in homes for heating and cooking. Coal has been the dominant source of emissions since the Industrial Revolution, and combustion of coal releases the most CO_2 per unit of energy, making it the dirtiest fuel. Oil is used in the transportation sector, where fuels such as gasoline and diesel are burned in internal combustion engines.

In 2019, 85.5% of global CO_2 emissions came from fossil fuels and industry.

The many uses of oil
Crude oil extracted from the ground must be distilled (separated) into usable forms at high temperatures. Smaller molecules, such as gasoline, have a lower boiling point and condense high up in the distillation column.

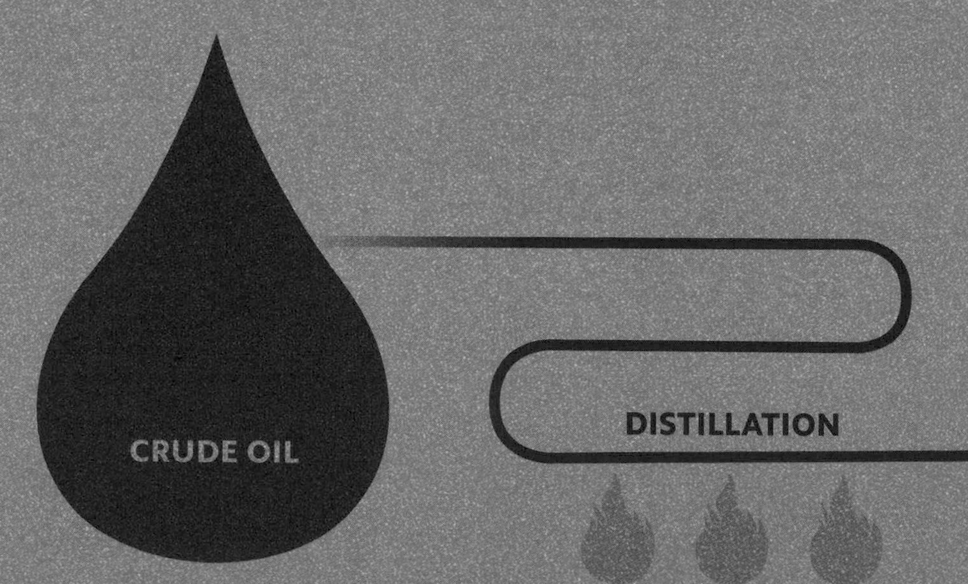

< 77°F (< 25°C)	**LIQUID PETROLEUM GAS** (USED FOR HEATING AND COOKING)
77–140°F (25–60°C)	**GASOLINE** (FUEL FOR ROAD VEHICLES)
140–356°F (60–180°C)	**NAPHTHA** (USED TO MAKE PLASTICS)
356-428°F (180–220°C)	**KEROSENE** (FUEL FOR AIRPLANES)
428–482°F (220–250°C)	**DIESEL** (FUEL FOR LARGER VEHICLES)
482–572°F (250–300°C)	**FUEL OIL** (FUEL FOR TANKERS OR GENERATING ELECTRICITY)
572–662°F (300–350°C)	**LUBRICATING OILS** (MOTOR OIL)
> 662°F (> 350°C)	**BITUMEN** (ASPHALT, USED FOR ROAD SURFACES)

DISTILLATION COLUMN

SUN

ATMOSPHERE

CARBON IN THE AIR

On average, CO_2 stays in the atmosphere between 300 and 1,000 years.

NOTHING NEW UNDER THE SUN

On a short-term timescale, carbon is naturally cycled between stores, or sinks, including the atmosphere, ocean, and biosphere. On much longer timescales, carbon is removed from this cycle by carbon stores, such as the fossil fuels that formed over millions of years. Human activity is adding carbon from these long-term carbon stores into the atmosphere. Although absorption of carbon from the atmosphere has increased, atmospheric carbon is still rising.

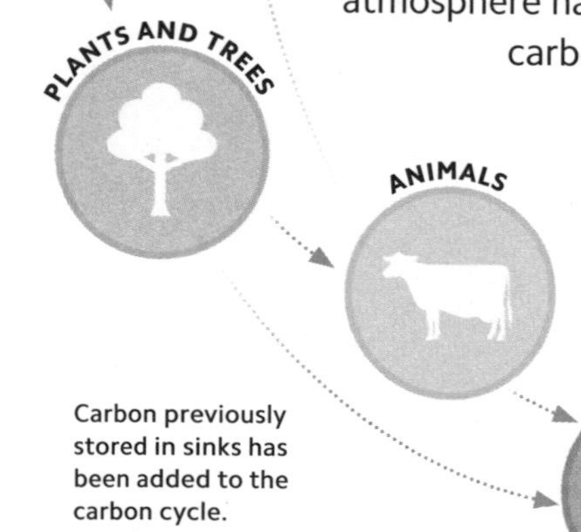

ROCKS

OCEANS

HUMAN ACTIVITY

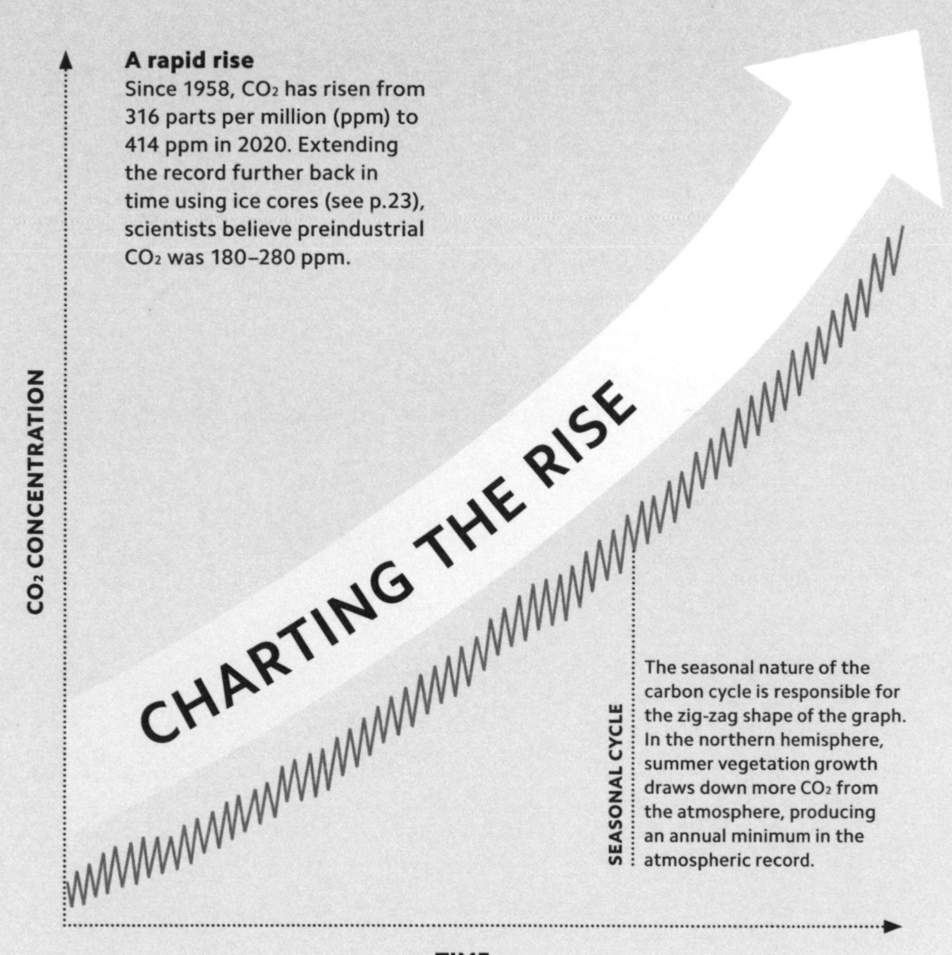

A rapid rise
Since 1958, CO_2 has risen from 316 parts per million (ppm) to 414 ppm in 2020. Extending the record further back in time using ice cores (see p.23), scientists believe preindustrial CO_2 was 180–280 ppm.
CHARTING THE RISE
CO_2 CONCENTRATION
SEASONAL CYCLE
The seasonal nature of the carbon cycle is responsible for the zig-zag shape of the graph. In the northern hemisphere, summer vegetation growth draws down more CO_2 from the atmosphere, producing an annual minimum in the atmospheric record.
TIME

Scientists began measuring the amount of CO_2 in the atmosphere in 1958, at a high-altitude station in Mauna Loa, Hawaii. The record of CO_2 taken at this station has shown the continual increase in atmospheric CO_2 since then, in a figure known as the Keeling Curve, named after American scientist Charles Keeling. Since 2000, atmospheric CO_2 has been growing at a faster rate than ever before, as human emissions have been increasing.

A STRONG CONSENSUS

Despite the global spread of misinformation and climate-change denialism, scientists agree that climate change is primarily human-driven. They use climate models (see pp.24–25) and run studies to accurately estimate the degree of global warming attributable to human activity. More significantly, real-time observation of the effects of global warming, such as melting ice (see p.88), drying deserts, and shifting ocean currents (see pp.18–19) has also contributed to this scientific consensus.

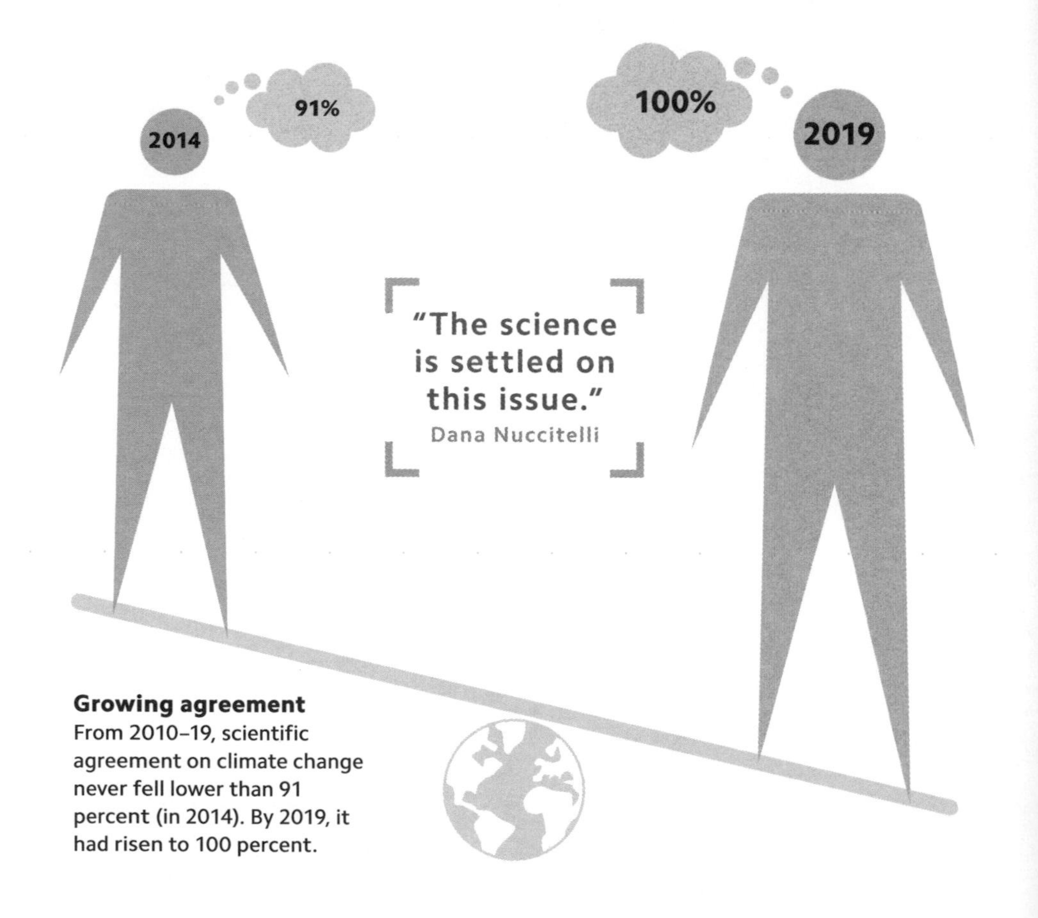

Growing agreement
From 2010–19, scientific agreement on climate change never fell lower than 91 percent (in 2014). By 2019, it had risen to 100 percent.

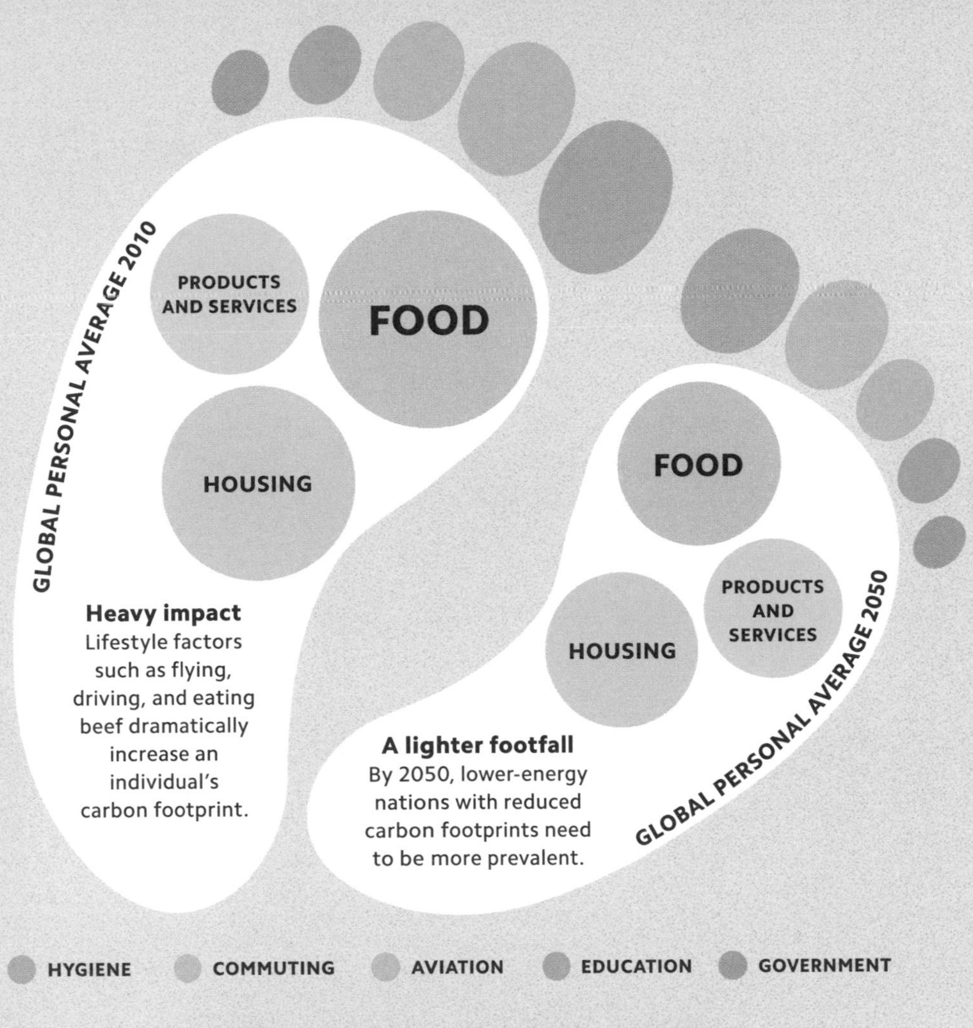

MIND YOUR STEP

All societies rely on industries, services, or activities that cause the release of emissions. The amount of greenhouse gases emitted into the atmosphere by an individual, business, or product is called a carbon footprint. The carbon footprint of those in higher-income nations is often larger than that of those in lower-income nations. Any individual or business can take steps to reduce their carbon footprint, and doing so is an effective method of fighting climate change.

WHAT'S YOUR BUDGET?

There is a finite amount of carbon left that can be emitted before global temperature rise exceeds a certain limit. This amount is known as the carbon budget. While carbon budgets can be set for any temperature rise limit, most hinge on the 2.7°F (1.5°C) limit outlined in the Paris Agreement. Unless emissions reach zero, our carbon budget will lessen every year. The more carbon that we emit, the more of the budget we use up, and the worse the effects on our climate.

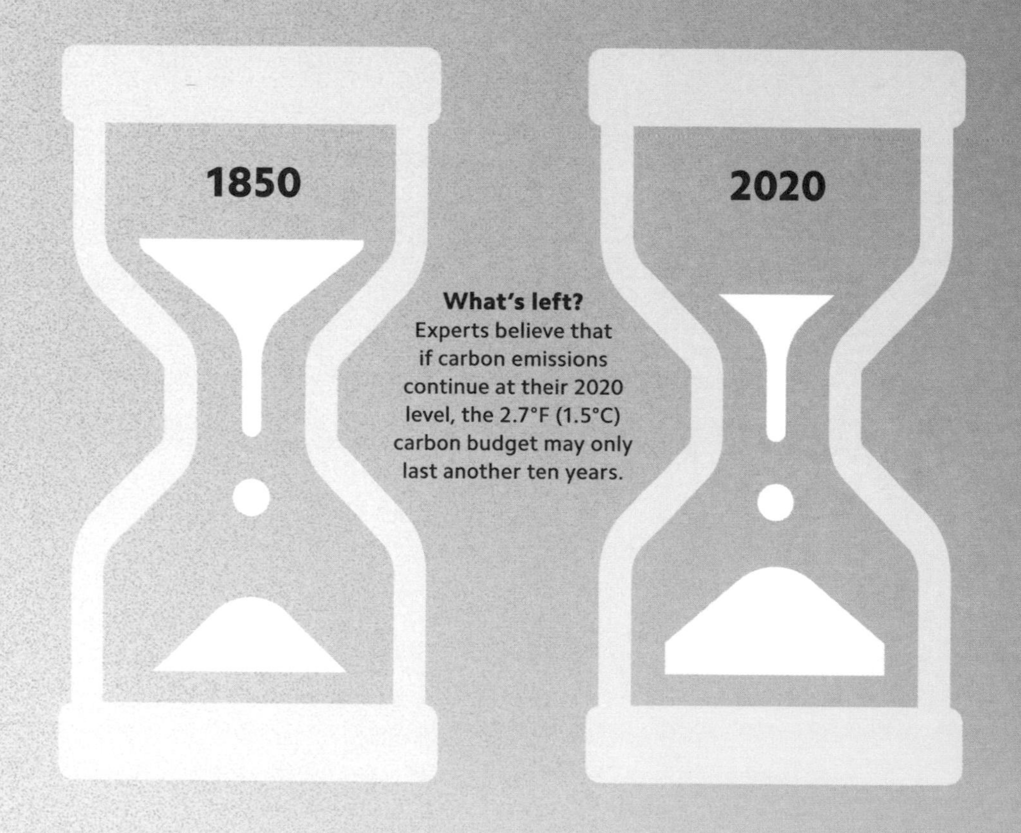

What's left?
Experts believe that if carbon emissions continue at their 2020 level, the 2.7°F (1.5°C) carbon budget may only last another ten years.

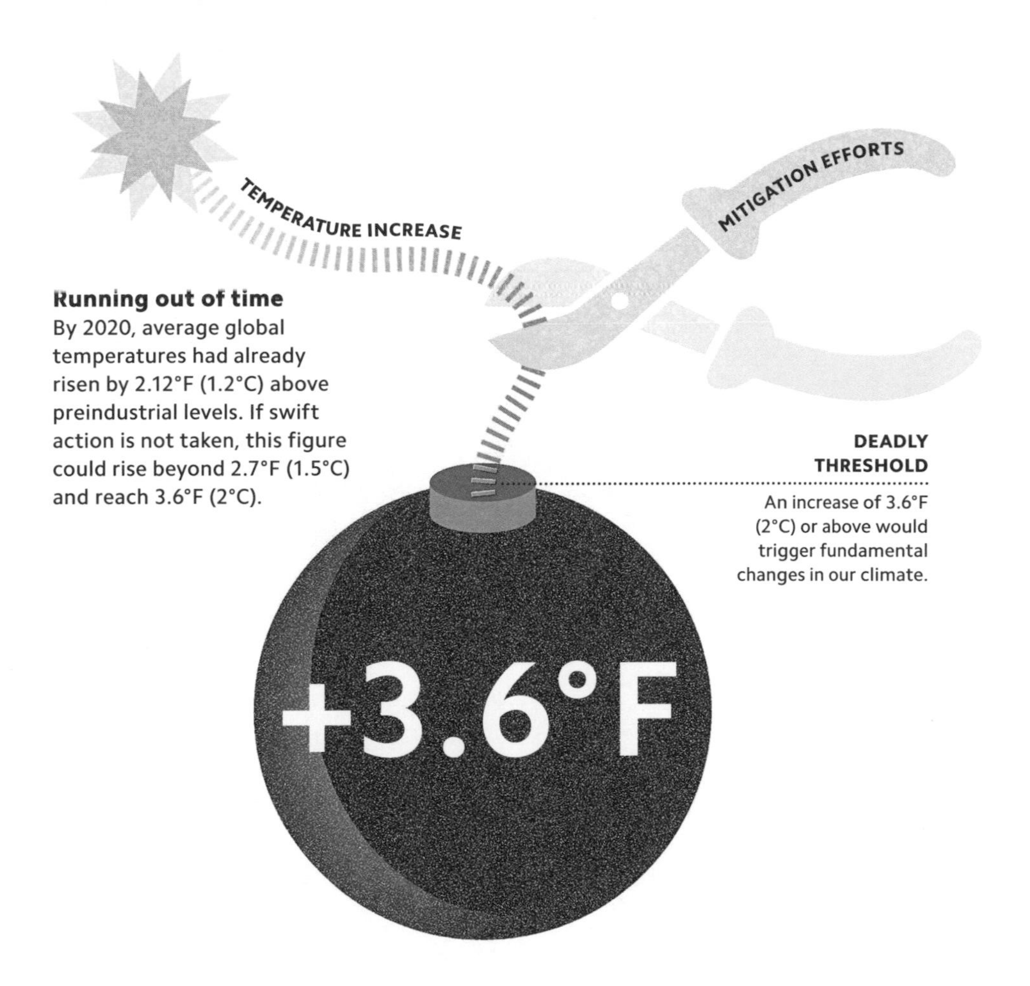

TEMPERATURE INCREASE
MITIGATION EFFORTS
Running out of time
By 2020, average global temperatures had already risen by 2.12°F (1.2°C) above preindustrial levels. If swift action is not taken, this figure could rise beyond 2.7°F (1.5°C) and reach 3.6°F (2°C).
DEADLY THRESHOLD
An increase of 3.6°F (2°C) or above would trigger fundamental changes in our climate.
+3.6°F

A DEGREE OF CONCERN

International efforts aiming to limit global warming to 2.7°F (1.5°C) stress that a rise of 3.6°F (2°C) carries significantly greater risks for people and nature. At 3.6°F (2°C), coral reefs, already greatly impacted by climate change (see p.104), would almost entirely vanish, while extreme weather events would become more common and more intense. Every fraction of a degree of warming that is prevented reduces these and other effects of climate change, such as wildfires (see p.84), heatwaves, and flooding.

PHASE OUT COAL

HALT DEFORESTATION

SHIFT TO ELECTRIC VEHICLES

BUILD RENEWABLE ENERGY

RETROFIT HOMES

DECARBONIZE ELECTRICITY

INCREASE PUBLIC TRANSPORTATION

CHANGING OUR TECHNOLOGIES
Upscaling existing renewable energy technologies, along with phasing out coal, is a crucial step in any serious bid to reach net zero emissions.

CONVERTING OUR HOMES
The technology to reduce a home's energy demand by 80 percent now exists (see pp.126–27). Implementing it, however, requires factory-produced, large-scale solutions.

A greener future
Generating electricity with wind and solar power instead of coal and gas is one of a number of strategies governments have begun to implement. Removing carbon emissions from the production of industrial materials, however, is harder to achieve.

> **As of 2020, just six of 195 countries, including Sweden and the UK, have set a legally binding net-zero target.**

To stablize the average global temperature, carbon emissions from human activity need to reach net zero. A net-zero target aims to add no more carbon to the atmosphere, which means that if there are any emissions, the same amount of carbon needs to be removed to offset them. The creation of carbon sinks—natural environments, such as a forest, that absorb carbon—is one way of doing this. To cut emissions in the first place, the sources—in particular fossil fuels—need to be replaced with carbon-free alternatives. Individuals and businesses can also help by making reductions to their carbon footprint (see p.33), such as replacing private trips with public transportation alternatives.

DECARBONIZE MATERIALS

CUT FOOD WASTE

STEPS TO NET ZERO

P O P U L

ATION

The global human population took about 2,700 years to grow from 150 million to 500 million, a figure reached in the mid-17th century. It has boomed dramatically since and is expected to pass 9.7 billion by 2050. Advances in agriculture were followed by technological and medical innovations that slashed infant mortality, improved public health, and increased life expectancy. The almost 16-fold increase in the world population since the 1650s has greatly amplified human impact on the environment.

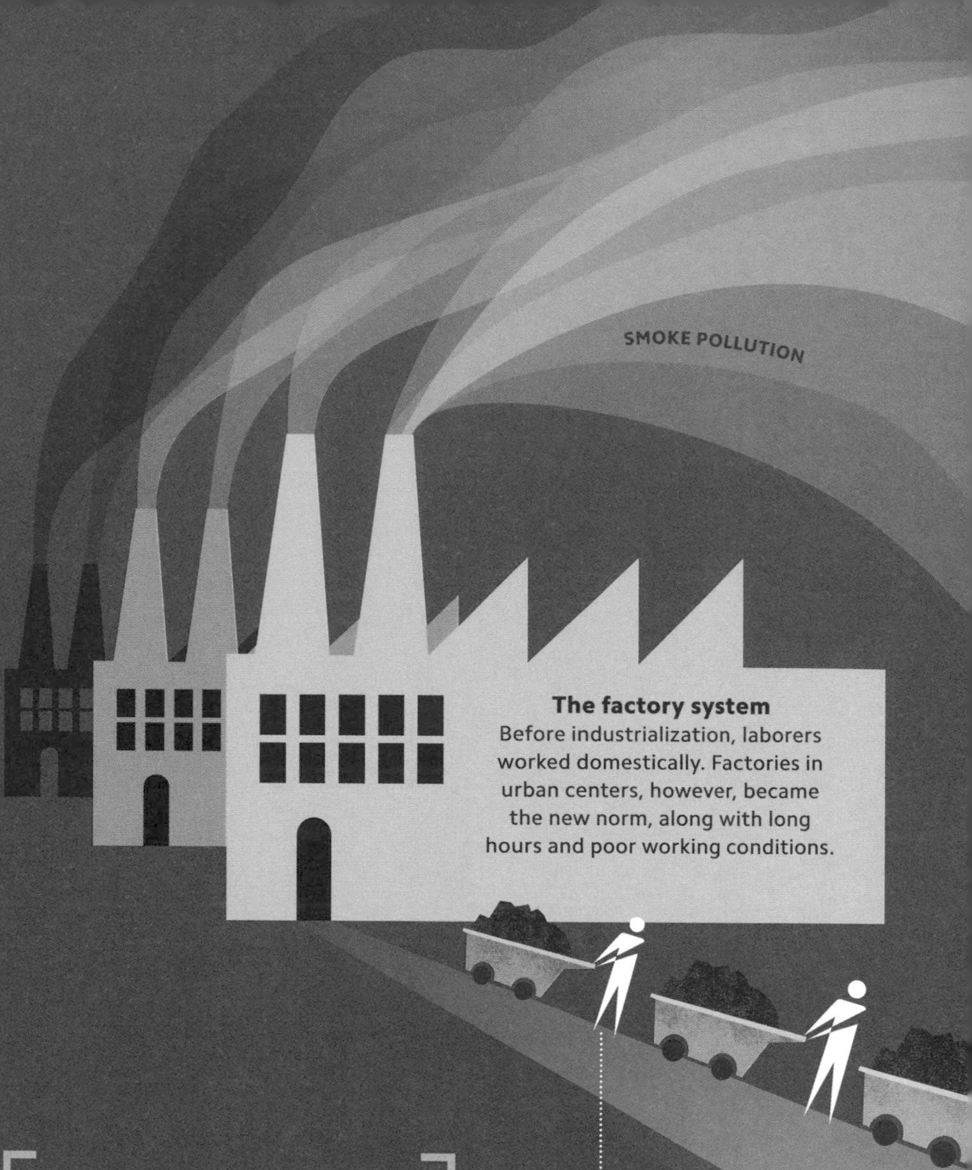

SMOKE POLLUTION

The factory system
Before industrialization, laborers worked domestically. Factories in urban centers, however, became the new norm, along with long hours and poor working conditions.

RISING LABOR
As the scale of mass production exploded, so did the demand for unskilled workers, who flocked to factory jobs in towns and cities.

Since the Industrial Revolution, human actions have increased CO_2 concentration in our atmosphere by 48%.

HOW IT ALL BEGAN

From the mid-18th century, the US and Europe gradually shifted from an agricultural economy to one led by urban industry and mass production. This revolution, as it spread across the world, was largely fueled by the burning of coal in factory systems dealing with iron and steel, and the rise of the steam engine, which, from the mid-19th century, became the primary source of power for many nations. Later, technologies such as the internal-combustion engine became widespread. In the modern era, industrial growth continues to take a heavy toll on our planet.

INTO THE ATMOSPHERE

New industrial centers hosting vast working populations and fossil-fuel-based technological innovations all saw global air pollution skyrocket.

THE SHAPE OF THINGS TO COME

Before the 1970s, short life spans and high mortality rates meant that under-5s were the most numerous members of society globally. Since then, lower birth rates and improved health care have resulted in an increasingly aging population. Older age groups are more susceptible to some effects of climate change, such as heatwaves. In addition, these groups often have a greater carbon footprint than their younger counterparts (see pp.44–45).

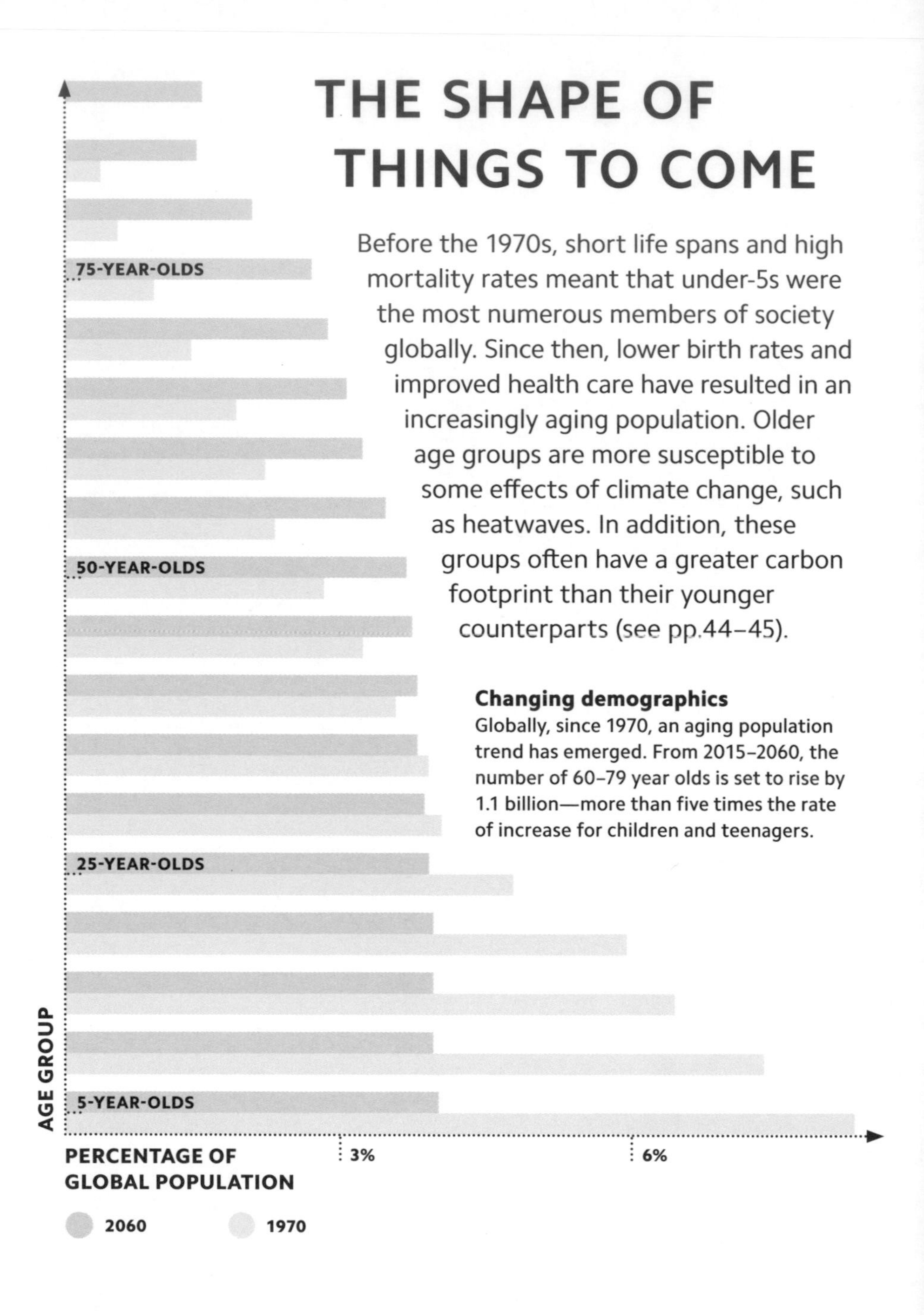

Changing demographics
Globally, since 1970, an aging population trend has emerged. From 2015–2060, the number of 60–79 year olds is set to rise by 1.1 billion—more than five times the rate of increase for children and teenagers.

New to the city
Megacities are economic centers. Job opportunities draw in new residents, who face heavy air pollution, strained public services, and often cramped living.

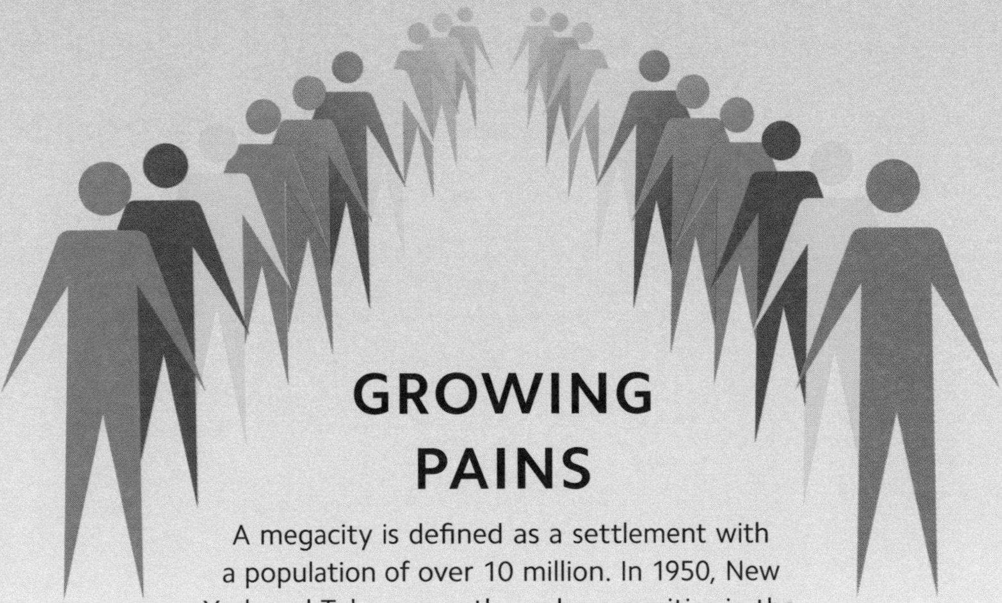

GROWING PAINS

A megacity is defined as a settlement with a population of over 10 million. In 1950, New York and Tokyo were the only megacities in the world. Today, more than 30 exist, and more than 40 are expected by 2030. When rapid urban growth goes unplanned, it can outpace infrastructure and services, leading to failures in emissions and pollution control, breakdowns in public health services, and rising amounts of solid and industrial waste. Many megacities are major hotspots for greenhouse gas emissions, impacting far beyond their city limits.

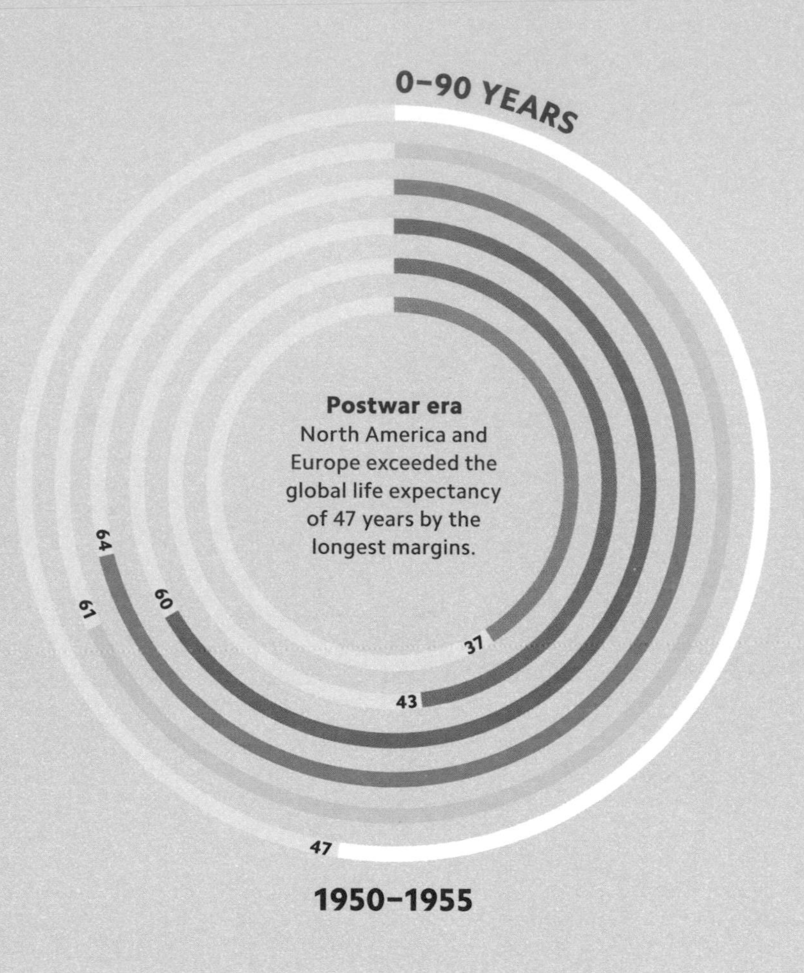
0–90 YEARS
Postwar era
North America and Europe exceeded the global life expectancy of 47 years by the longest margins.
64
61
60
37
43
47
1950–1955

A RIPE OLD AGE

In 1800, global life expectancy was less than 35 years. By 2020, it had more than doubled, reaching 73 years. This is largely due to improved sanitation, health education, and medicine. There is great disparity between the life expectancies of higher-income and lower-income nations. In all parts of the world, people living longer result in greater individual carbon footprints, and added strain on natural resources.

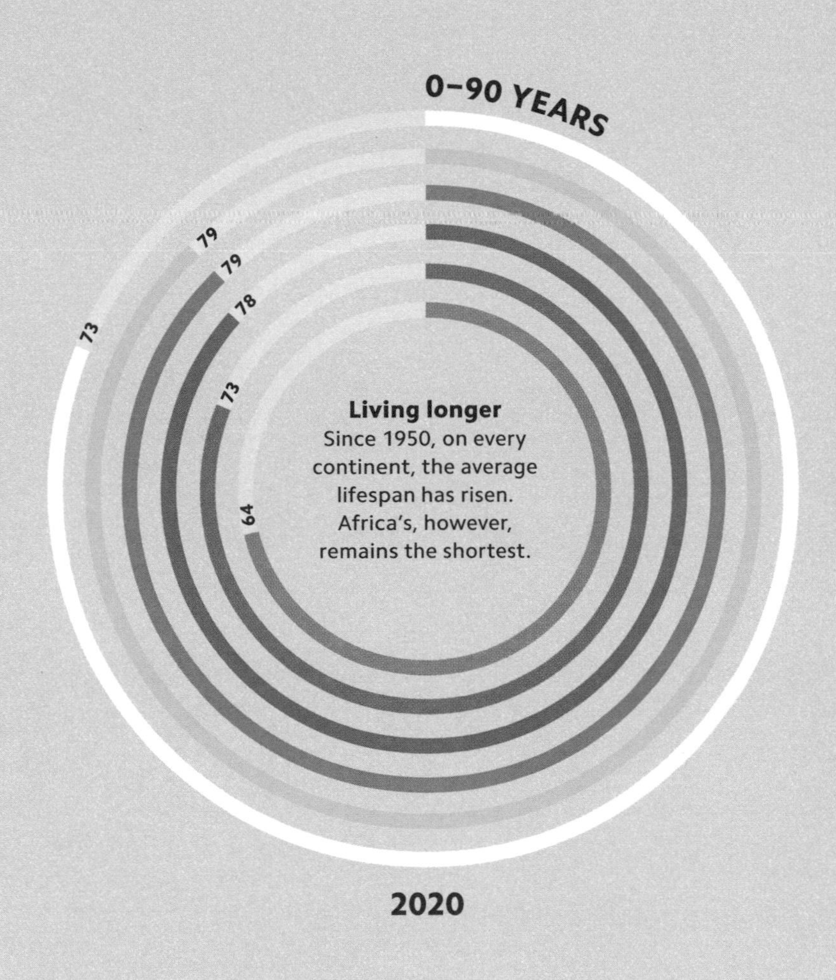

0–90 YEARS
79
79
78
73
73
64
Living longer
Since 1950, on every continent, the average lifespan has risen. Africa's, however, remains the shortest.
2020

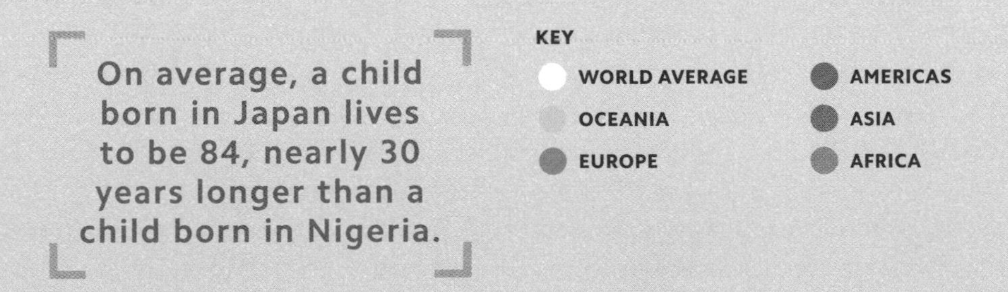

On average, a child born in Japan lives to be 84, nearly 30 years longer than a child born in Nigeria.
KEY
WORLD AVERAGE
OCEANIA
EUROPE
AMERICAS
ASIA
AFRICA

FOOD A
RESOU

N D
R C E S

Boosted by higher crop yields in cereal agriculture from the 1960s onward, global food production has managed to keep pace with a booming human population, but at a cost to other resources and the environment. Farming occupies approximately half of the planet's habitable land and consumes 70 percent of all freshwater withdrawals. Despite the rise in food production, poverty, waste, conflict, and inequities cause many millions of people to go hungry and undernourished every year.

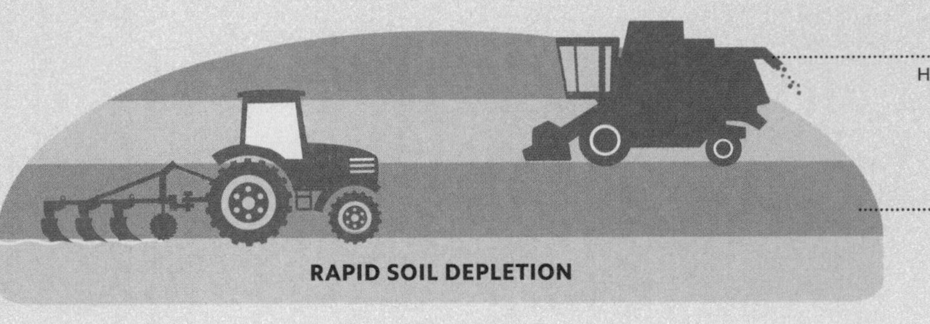

MECHANIZATION

Heavy mechanization of the growing and harvesting process results in fossil fuel use and pollution.

MONOCULTURES

Repeatedly growing a single crop in one area lowers biodiversity and leads to rapid soil-nutrient depletion.

THE COST OF FOOD

Heavy investment in mechanization, labor, pesticides, and fertilizer results in higher crop yields per unit area than other farming systems. Adherents claim it ensures a consistently high volume of affordable food to satisfy growing world demand. Critics, though, highlight environmental impacts, including land clearance, causing a loss of biodiversity, pollution, monoculture crops degrading soil quality, and the cruelty and potential health risks of factory-farmed livestock, packed or caged inside giant sheds and forced to grow at unnatural rates.

Animals are densely crammed into small enclosures and may be fed grain with additives such as antibiotics and hormones to fight the resulting infections and speed growth.

FACTORY FARMING

LAND GRAB

The need for new land to grow more food has led to large-scale land clearances, which has severe impacts on wildlife and local communities.

High-yield agriculture

Intensive farming has led to higher yields but at significant environmental cost. Its impacts are multilayered, including high emissions from farm machinery, reduced biodiversity, and pollution via pesticide and fertilizer runoff (see pp.52–53).

> "The industrial food system presents a barrier to realizing the potential climate benefits in agriculture"
>
> Laura Lengnick

MAKE IT GROW

In 1913, the Haber-Bosch process began the industrial-scale fixation (transformation) of atmospheric nitrogen into substances used in artificial fertilizers to enrich soils and support heavier crop yields. The world's farms now rely on more than 220 million tons (200 million tonnes) of artificial fertilizers annually, but this comes at a cost.

The Haber-Bosch process consumes more than one percent of the world's energy, while fertilizer nutrients can damage aquatic ecosystems and lead to increased emissions of the greenhouse gas nitrous oxide (N_2O).

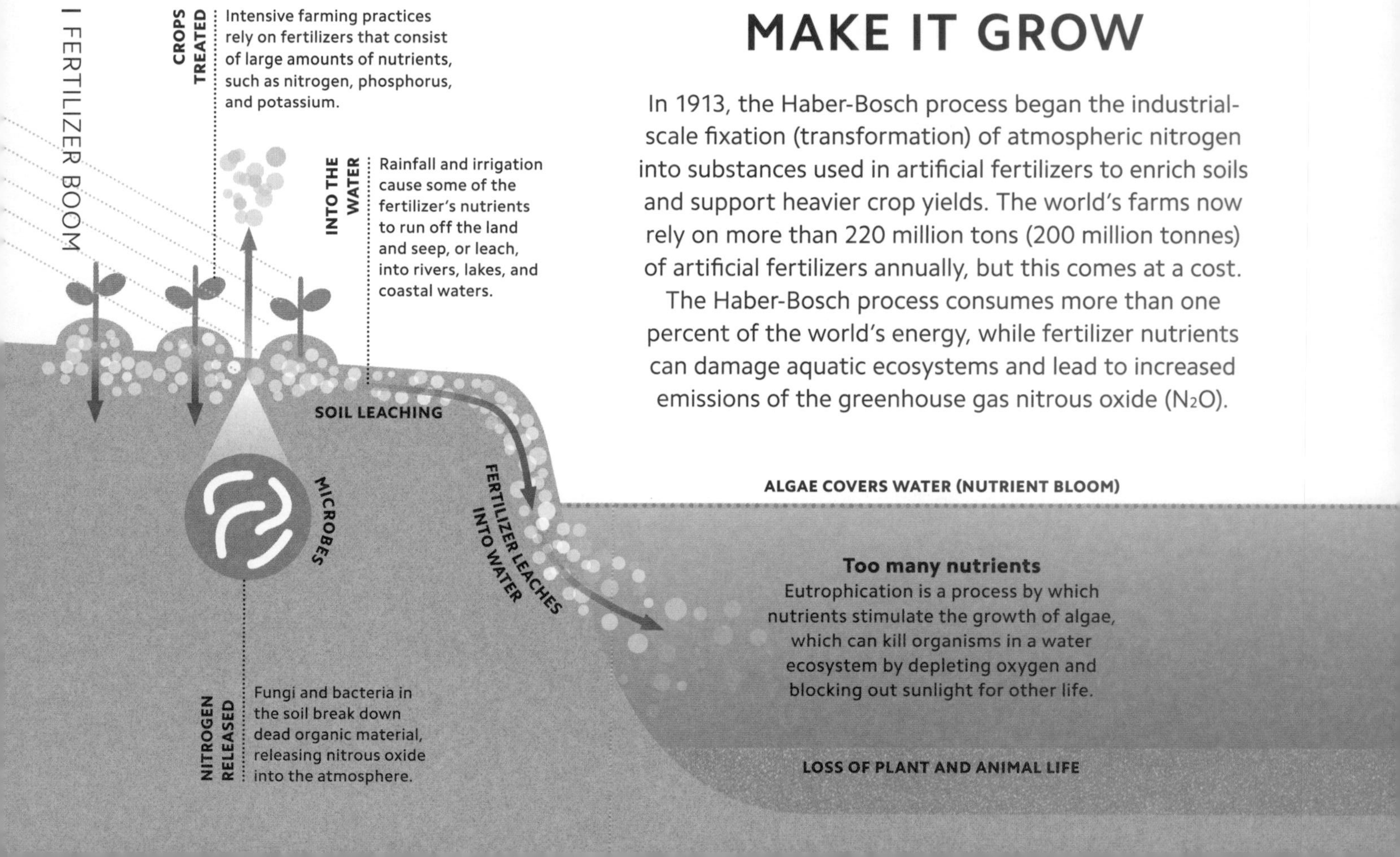

CROPS TREATED Intensive farming practices rely on fertilizers that consist of large amounts of nutrients, such as nitrogen, phosphorus, and potassium.

INTO THE WATER Rainfall and irrigation cause some of the fertilizer's nutrients to run off the land and seep, or leach, into rivers, lakes, and coastal waters.

SOIL LEACHING

MICROBES

FERTILIZER LEACHES INTO WATER

NITROGEN RELEASED Fungi and bacteria in the soil break down dead organic material, releasing nitrous oxide into the atmosphere.

ALGAE COVERS WATER (NUTRIENT BLOOM)

Too many nutrients
Eutrophication is a process by which nutrients stimulate the growth of algae, which can kill organisms in a water ecosystem by depleting oxygen and blocking out sunlight for other life.

LOSS OF PLANT AND ANIMAL LIFE

TOXINS IN OUR FOOD CHAIN

Hailed as chemical miracle workers that greatly increased crop yields, many pesticides have proven toxic to more than the insect, fungi, and plant pests they were designed to control or eradicate. A fifty-fold rise in their use since 1950 has been accompanied by their accumulation in soils, water sources, and food chains where biomagnification sees them reduce the healthy numbers of top predators. Pesticides can also reduce populations of harmless or helpful species, damaging biodiversity (see p.86).

CONCENTRATED CONTAMINATION

Animals at the top of the food chain will consume the greatest concentration of biomagnified toxins.

Biomagnification
Larger creatures eat a lot of smaller organisms to fulfil their dietary needs and small amounts of toxins become concentrated higher up the food chain.

CUTTING DOWN DEFENSES

According to the United Nations' Food and Agriculture Organization, 1.6 million square miles (4.2 million square km) of forest—an area 6.5 times the size of France—has been lost since 1990. The key driver for this is land clearance for agricultural expansion. This alarming loss of rich habitats not only threatens biodiversity and reduces the role of forests as vital carbon sinks, but also removes the soil erosion cover and flood protection that trees and their roots provide.

Rapid decline
The increased need for land and resources led to large-scale deforestation from the start of the Industrial Revolution (see pp.40–41). The net loss of temperate forest peaked in the 1990s.

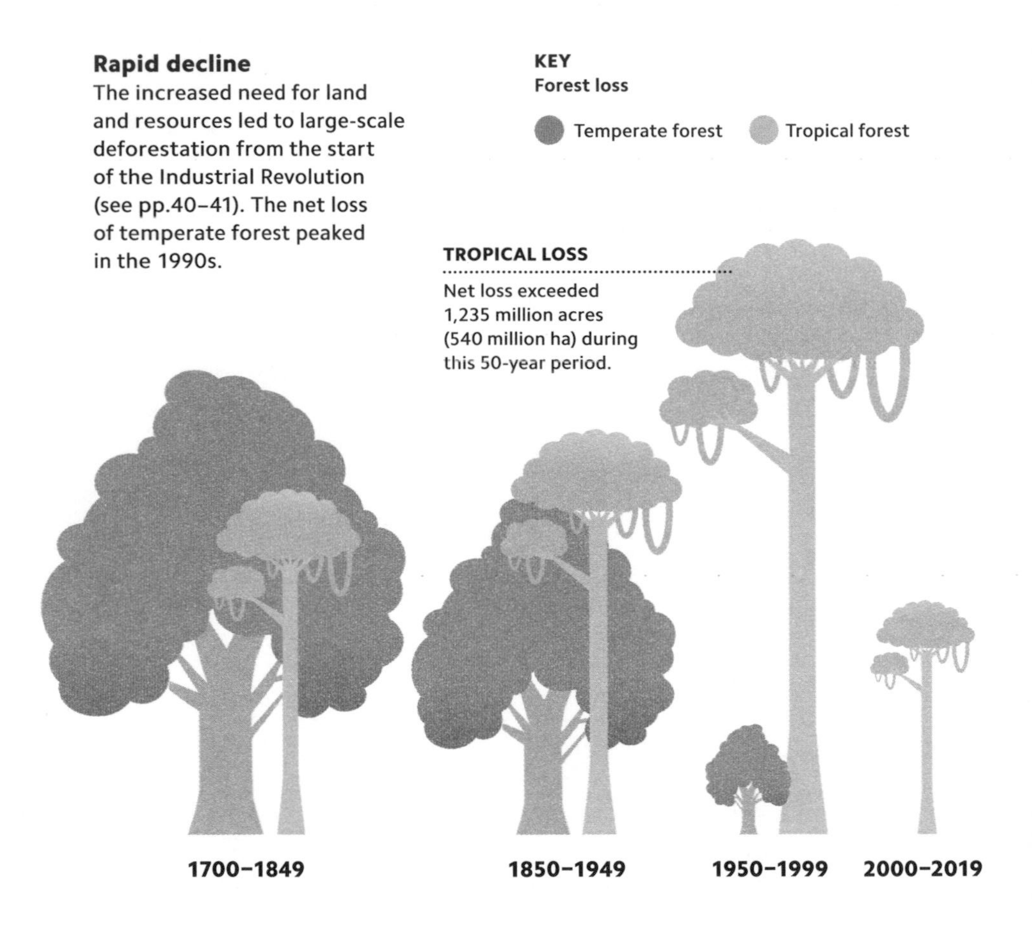

PLENTY LESS FISH IN THE SEA

Caught out
Nearly a third of fish stocks have been overfished, which shifts the pressure onto stocks already being fished to capacity.

61% FISH STOCKS FULLY FISHED

29% FISH STOCKS OVERFISHED

10% FISH STOCKS UNDERFISHED

A soaring global demand for fish and seafood has been met by plundering the oceans, with damage done to fish stocks and ecosystems. The number of fish stocks endangered by overfishing has tripled in the past 50 years as breeding populations become too depleted to recover and food webs are degraded as a result. Indiscriminate trawling and fishing also kills over 33 million tons of unwanted by-catch (unwanted animals caught and killed in the process) annually. A rise in aquaculture (farming), is meeting some of the demand for seafood.

TAKING OVER

Whether through trade, transport, or driven by climate change, many species are spreading outside their native range and wreaking havoc on organisms and ecosystems in their new homes. When free of the natural checks and predation found in their native environment, burgeoning populations of invasive species may soar in number and outcompete indigenous species for resources. They may force native species into endangerment or extinction, lowering biodiversity and disrupting the delicate ecosystem balance.

MOUNTAIN PINE BEETLE

These wood-boring insects kill off pine trees and have affected millions of acres of North American forest. Warmer winters have allowed the beetle to extend its range northward.

> "A warmer climate could allow some invaders to spread farther."
> Richard Preston

SEAWEED

Outside its native Pacific Ocean, the seaweed *Caulerpa taxifolia* is a highly invasive species. Whether it has a negative impact or not is debated.

LIONFISH

A voracious eater, a single lionfish, as an invasive species, can consume half of the fish in a coral reef habitat in just six weeks.

CANE TOAD

In 1934–35, 2,400 South American cane toads were set free in Australia as pest controllers on sugar cane plantations. They now number as many as 1–1.5 billion.

1.4 billion tons (1.3 billon tonnes) of food is wasted each year.

NO NUTRITION

As much as one-third of all food and the resources that go into its production is wasted, according to estimates from the UN Food and Agriculture Organization (FAO). Waste occurs at every stage, including initial production, sorting, shipping, retail, and household consumption. About 40 percent of all food waste in developed nations occurs at the retail stage. Food waste results not just in squandered resources, it also generates significant greenhouse gas emissions, especially methane from rotting food.

LAND WASTE

An estimated 28 percent of the world's farmland is used to produce lost and wasted food.

WATER RESOURCES

Food wastage uses an estimated 60 cubic miles (250 cubic km) of water every year.

CARBON COST

The estimated carbon footprint of wasted food is 3.6 billion tons (3.3 billion tonnes) of CO_2-equivalent greenhouse gases released into the atmosphere.

FARMLAND

WATER WASTE

CARBON FOOTPRINT

WATER FOOTPRINT

Direct water consumption, which is 4–143 gallons (15–540 litres) per capita daily, is just the tip of the iceberg; everything bought and consumed uses water along the way. Water footprints measure both the water consumed and polluted. They can be calculated for individuals, processes, or the full production cycle of products—from supply chain to end-user delivery. The personal water footprint for those living in developed nations may run into thousands of gallons per day.

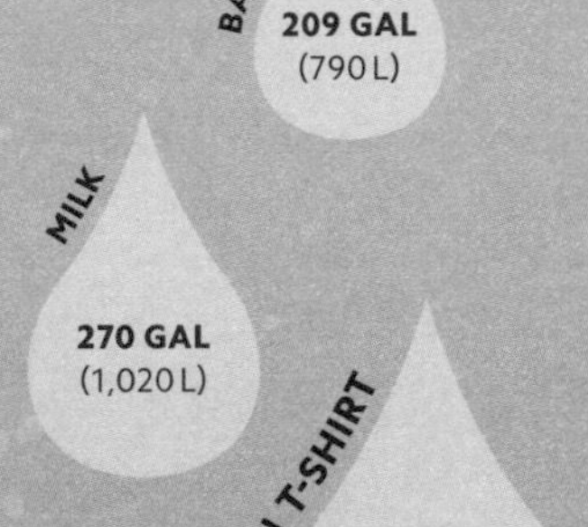

Big footprints
The average water footprint for some common food items are shown here in gallons per pound produced. The average footprint of one cotton T-shirt depends on the fabric's thickness.

R I S I N
C O N S U

G
MPTION

All forms of consumption can contribute to climate change. The contribution through consumption is greater in richer countries, but economic development is driving up the consumption of many poorer countries. As a result, the global average emissions per person—currently about 5.3 tons (4.8 tonnes) per person—has doubled since 1950, and is continuing to climb. While some sectors are cutting carbon emission, everything from gadgets to fashion has become more disposable and more polluting. Many countries appear to have cut their consumption, but have merely shifted the production of their products to other countries.

CHEAP AND DIRTY ENERGY

Coal has provided cheap power since the industrial revolution (see pp.40–41), but that energy has come at a cost. Burning coal is both dirty and deadly; it produces 50 percent more CO_2 than burning gas and its pollution is estimated to cost at least three times more lives than any other energy source. Coal use has declined dramatically in many more economically developed countries (MEDCs) and some coal-generating countries are already planning a complete phaseout.

OUTGOING FUMES

POLLUTED FUMES

In a modern plant, waste gases are treated to clean them, though for a long time, harmful emissions were simply released into the atmosphere.

COAL PARTICLES ARE PARTIALLY FILTERED

COAL SUPPLIED TO FURNACE

> "Coal-fired power stations are death factories."
>
> James Hansen

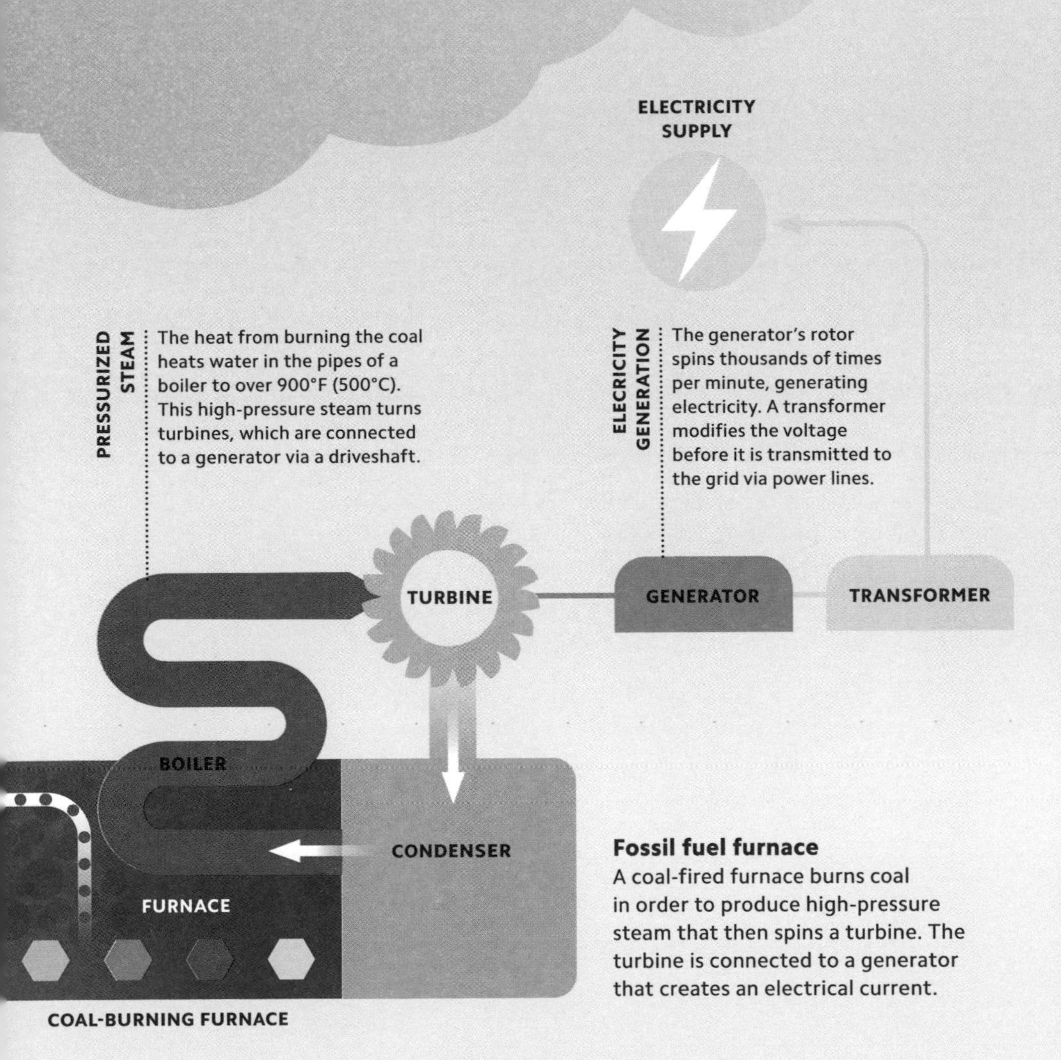

ELECTRICITY SUPPLY
PRESSURIZED STEAM
The heat from burning the coal heats water in the pipes of a boiler to over 900°F (500°C). This high-pressure steam turns turbines, which are connected to a generator via a driveshaft.
ELECRICITY GENERATION
The generator's rotor spins thousands of times per minute, generating electricity. A transformer modifies the voltage before it is transmitted to the grid via power lines.
TURBINE
GENERATOR
TRANSFORMER
BOILER
CONDENSER
FURNACE
COAL-BURNING FURNACE

Fossil fuel furnace
A coal-fired furnace burns coal in order to produce high-pressure steam that then spins a turbine. The turbine is connected to a generator that creates an electrical current.

ROAD TO DESTRUCTION

Vehicles are key drivers of climate change, with road transportation accounting for more than 10 percent of global greenhouse gas emissions. Of all road vehicles, large cars, such as Sport Utility Vehicles (SUVs), are especially damaging to the environment, and show no sign of waning in popularity, currently making up 39 percent of global car sales compared to 17 percent in 2010. Efforts by climate activists to reduce pollution and harmful emissions often include alternatives to short-trip car use, such as public transportation, cycling, and walking, all of which help to minimize unnecessary car trips.

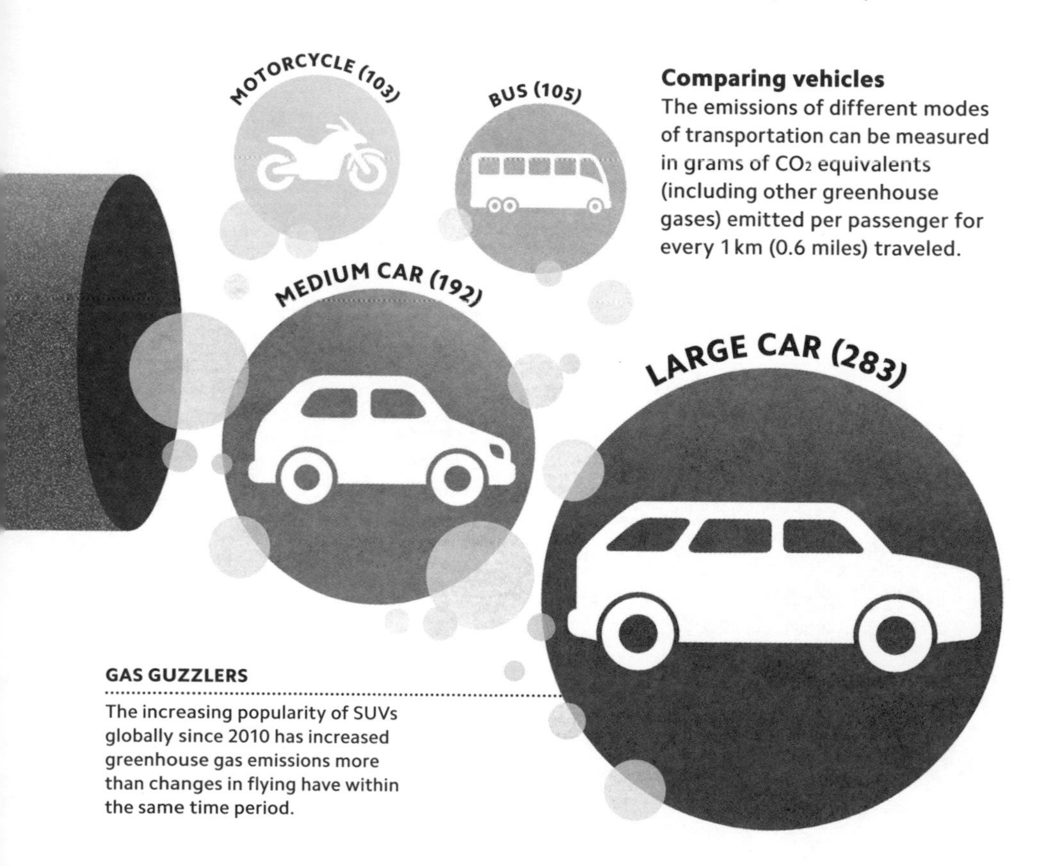

Comparing vehicles
The emissions of different modes of transportation can be measured in grams of CO_2 equivalents (including other greenhouse gases) emitted per passenger for every 1 km (0.6 miles) traveled.

GAS GUZZLERS

The increasing popularity of SUVs globally since 2010 has increased greenhouse gas emissions more than changes in flying have within the same time period.

Clouds of our making
Airplanes release vapor trails called contrails, which form cirrus clouds that can last minutes or hours. These clouds can trap heat rising from Earth, greatly adding to global warming.

The flight industry is responsible for about 2.4% of global CO_2 emissions.

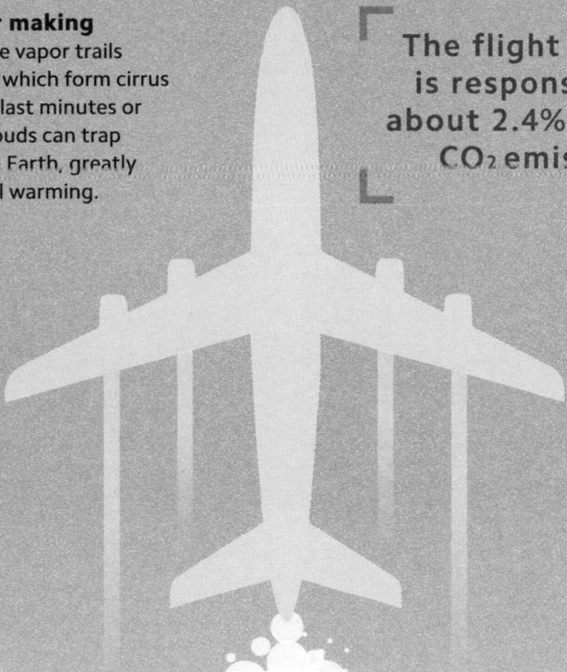

IN A TAILSPIN

Few transportation-related causes of climate change are as notorious as flying, and for good reason. Taking one long-haul flight produces more CO_2 emissions than many people's entire annual carbon footprint (see p.33). In addition to emitting CO_2, aircraft release other pollutants, which triple the overall warming effect of a trip. Until the start of the coronavirus pandemic in 2020, flying was increasing annually. If this trend returns and persists, emissions from aviation could use up a quarter of our 2.7°F (1.5°C) carbon budget by 2050.

HEAVY INDUSTRY

The production of metals, chemicals, and cement all depend on fossil fuels, and all result in heavy emissions. Producing 0.98 tons (one tonne) of steel, for example, generates an average of 1.87 tons (1.9 tonnes) of CO_2. This problem shows no signs of slowing down, as demand for steel and cement has already more than doubled this millennium. So far, only a few costly low-carbon alternatives exist. For many heavy industries, the most viable option is thought to be hydrogen, put through a process known as steam methane reforming (SMR).

INDUSTRIAL PRODUCTION

41 percent of heavy industry CO_2 emissions come from burning fossil fuels to create heat, which is then used to produce materials such as steel, cement, and petrochemicals.

Averting disaster

Heavy industry plays a key role in the economies of many countries, making it difficult to effectively decarbonize. If high-temperature heat alternatives to fossil fuels are not developed and adopted, heavy industry will continue to play a major role in the destruction of our planet.

Heavy industry accounts for about 22 percent of global CO_2 emissions.

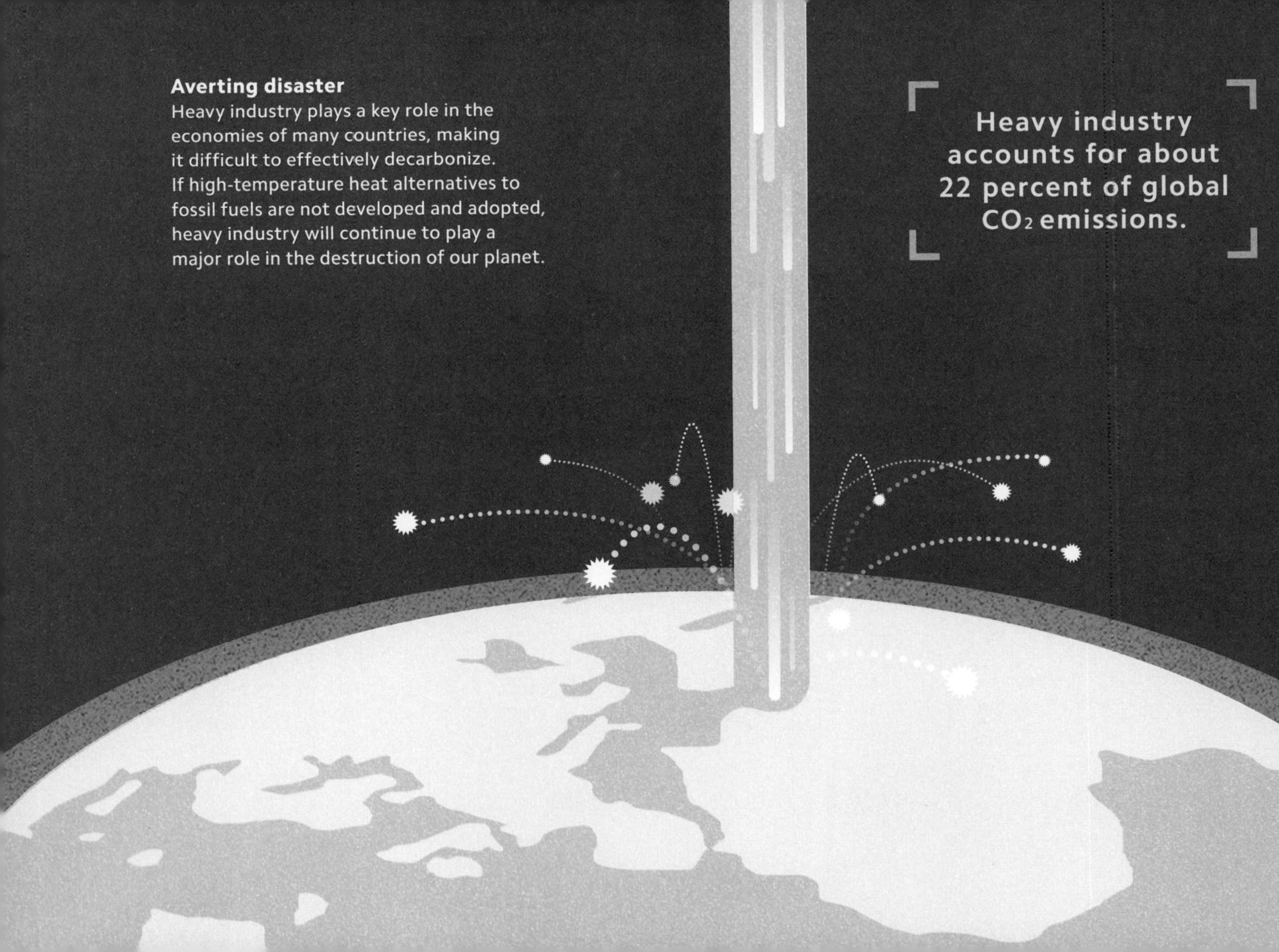

80 percent of all clothes sold in the EU are not recycled.

A damaging industry
In 2019, a UK government report found that the textile industry contributes more to climate change than aviation and shipping combined.

FASHION INDUSTRY
10%
OF ALL CO_2 PRODUCED

COTTON PRODUCTION
22.5% OF THE WORLD'S INSECTICIDES AND 10% OF PESTICIDES

THROWAWAY FASHION

Every year, 80–100 billion pieces of new clothing are purchased. This represents a 400 percent increase in the last 20 years. Much of this clothing is fast fashion—inexpensive, nondurable, and rapidly produced at high volume. Fast fashion consumes vast resources, generates significant greenhouse gas emissions, and produces clothes worn only briefly before they are considered disposable. Little is recycled; the vast majority is incinerated or buried in landfill, where polyester and other synthetic fibers can take centuries to degrade.

WASTEFUL WORLD

In 2018, annual municipal solid waste passed 1.97 billion tons (2 billion tonnes) globally. Only 13.5 percent of this was recycled. The rest was disposed of via dumping, landfill, and incineration. These methods harm ecosystems and generate pollution, including enormous greenhouse gas emissions—about 0.98 tons (one tonne) of CO_2 for every ton of waste. Single-use plastics (see p.105) make up a sizeable part of the waste stream. Their ubiquity, durability, and lack of biodegradability pose significant ecological problems.

THERMOPLASTICS

This group of plastics includes drinks bottles, bags, containers, trays, and food packaging film. Thermoplastics can be reheated and reshaped repeatedly, and are easy to recycle.

THERMOSETS

Thermosets include hot drink cups, cutlery, bottle caps, and microwave dishes. None of these items can be recycled easily.

Single-use society

Corporate reform and government action is key to fighting excess waste. Individuals, too, can take action, by recycling and reusing objects, and by not using single-use plastics.

EFFEC

ONTHE

ATMOS

T S

P H E R E

The greenhouse effect begins in the layers of gas that surround Earth. As more thermal energy is trapped within Earth's atmosphere by these gases, air temperatures increase. This is what is driving the increases in temperature seen around the world since the Industrial Revolution. A warmer atmosphere also affects the water cycle, with greater evaporation and water storage in the atmosphere increasing the load of tropical cyclones and making extreme rainfall events more frequent and intense. Burning fossil fuels has also produced physical pollution that directly causes respiratory disease.

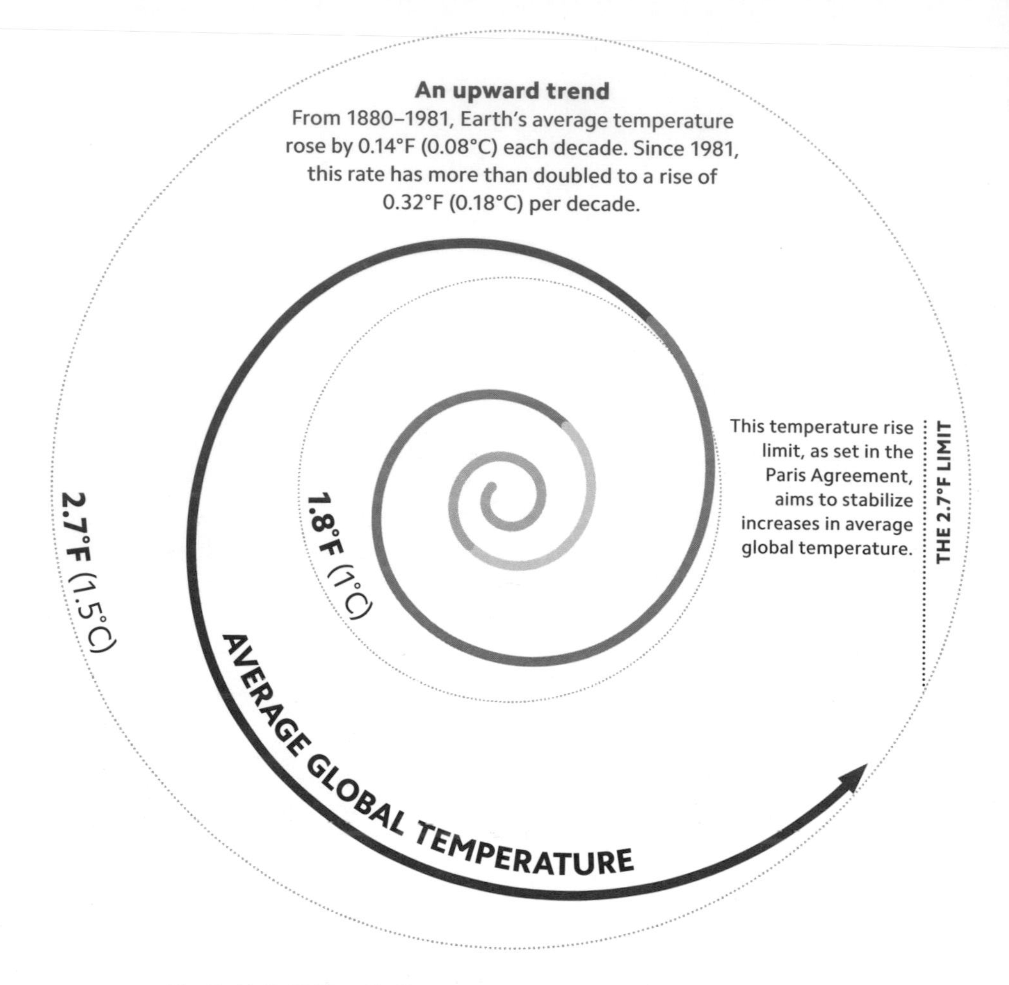

An upward trend
From 1880–1981, Earth's average temperature rose by 0.14°F (0.08°C) each decade. Since 1981, this rate has more than doubled to a rise of 0.32°F (0.18°C) per decade.
This temperature rise limit, as set in the Paris Agreement, aims to stabilize increases in average global temperature.
THE 2.7°F LIMIT
2.7°F (1.5°C)
1.8°F (1°C)
AVERAGE GLOBAL TEMPERATURE

CAN'T STAND THE HEAT

A rise in the average global temperature is the most commonly reported impact of climate change. At about 58.8°F (14.9°C) in 2020, the global average still seems low, but it is a useful metric by which to track the rate of climate change. Across the globe, temperature changes are not evenly distributed. For example, the Arctic is warming at three times the average global rate. However, virtually everywhere on the planet has increased in temperature since the Industrial Revolution (see pp.40–41).

As methods of measuring climate change have improved, a scientific consensus has emerged that the climate crisis is primarily caused by human activity (see p.32), discrediting alternative theories, such as natural variation in solar radiation. The scientific link between human-caused CO_2 emissions and global warming was first made in 1896. Today, the most advanced scientific models show that if it were not for the surge in greenhouse gas emissions that resulted from the Industrial Revolution, the average global temperature would have barely changed in the last 200 years.

WE CAUSED THE CRISIS

A very human problem

Scientists have developed a method of calculating how strong human influence is on extreme weather events (see p.75). This field is called extreme event attribution. Broadly, these studies have found that the intensity and scale of these events is often exacerbated by human activities.

HEATWAVES

EXTREME STORMS

FLOODING

WILDFIRES

Human health
Human societies are dependent on climate conditions, which influence factors such as food and freshwater supply. Many activities by humans affect the climate at both a global and local scale.

IT IS ALL CONNECTED

Climate health, human health, and ecosystem health are all inextricably linked, with each one directly influencing the others. Human activity, from farming to manufacture, influences the natural environment, directly impacting the atmosphere and the health of ecosystems in the ocean and on land. The resulting shifts in the biosphere have repercussions for all life, including humans. The natural links and feedback loops (see p.21) that exist between plant and animal communities and the physical environment mean that changes in one system have implications across the world. Climate change is resulting in changes happening too rapidly for systems to adapt and a weakend resilience of all components.

Ecosystem health

Trees and marine plankton are major regulators of atmospheric CO_2, and vegetation coverage has a major influence on the local climate and human activities. Habitat loss increases interactions between humans and animals, raising the risk of novel diseases jumping the species barrier.

ONE HEALTH

Climate health

A stable climate has a strong influence on the types of human and animal communities that inhabit regions all over the world.

> "The truth is: the natural world is changing. And we are totally dependent on that world."
> David Attenborough

UNCERTAINTY OF SEASONS

Many parts of the natural world respond to the seasonal variations in the climate. For some plants, flowering is triggered by temperature increases above certain thresholds. As temperatures warm and rainfall patterns shift, spring is, on average, arriving earlier in both the northern and southern hemisphere. This impacts human systems, in particular agriculture, and animal life. One key example is the Asian monsoon, where a change in the timing of the seasons influences the agricultural systems that feed billions of people.

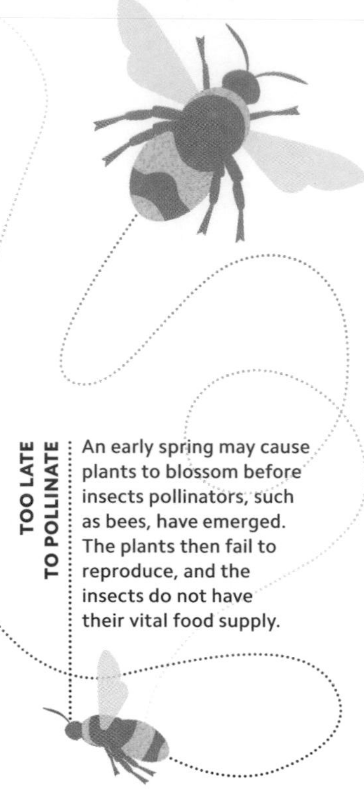

TOO LATE TO POLLINATE

An early spring may cause plants to blossom before insects pollinators, such as bees, have emerged. The plants then fail to reproduce, and the insects do not have their vital food supply.

SPRING

Seasons out of sync

Some animal and plant behaviors are triggered by seasonal changes in weather conditions. If climate change shifts the phasing of seasonal conditions, entire ecosystems could fall out of sync.

More extreme weather
The increased heat in the atmosphere changes patterns of evaporation and atmospheric circulation. This has been linked to unusual and extreme weather, such as intense, often deadly, heatwaves.

RAISING THE INTENSITY LEVELS

As the greenhouse effect (see p.13) increases, not only does the average temperature change, but the intensity and frequency of extreme climate events both increase. Longer heatwaves have set record temperatures across most parts of the world in recent years. Rainfall is also made more extreme, as warmer air can hold more moisture. This causes both more extreme drought, and also more intense rainfall, which can lead to devastating flooding.

EXTREME STORMS

Large tropical storms—called cyclones, hurricanes, or typhoons—have grown stronger since the 1980s. Taking their energy from warm ocean waters, these storms reach enormous wind speeds and can cause huge damage when they make landfall. Warmer ocean waters (see p.100) fuel stronger cyclones, and a greater proportion of these events now reach the higher sections of their classification scale, with wind speeds of more than 155 mph (250 kph).

Brewing storm

The largest storm systems form over equatorial ocean water. Warm, moist air rising from the ocean creates powerful circulating storms.

HEAVY CLOUDS

Once air reaches a high altitude, it cools and condenses to form thick cumulonimbus clouds, which produce heavy rainfall.

RISING AIR

The air above the ocean picks up the water's warmth and rises. This allows more air to rush in, which warms and rises, feeding an accelerating process that produces a cyclone.

STORM'S DIRECTION

RAIN

STRONG WINDS

Winds create bulges of water, which can cause high waves and flooding on land.

STORM SURGE

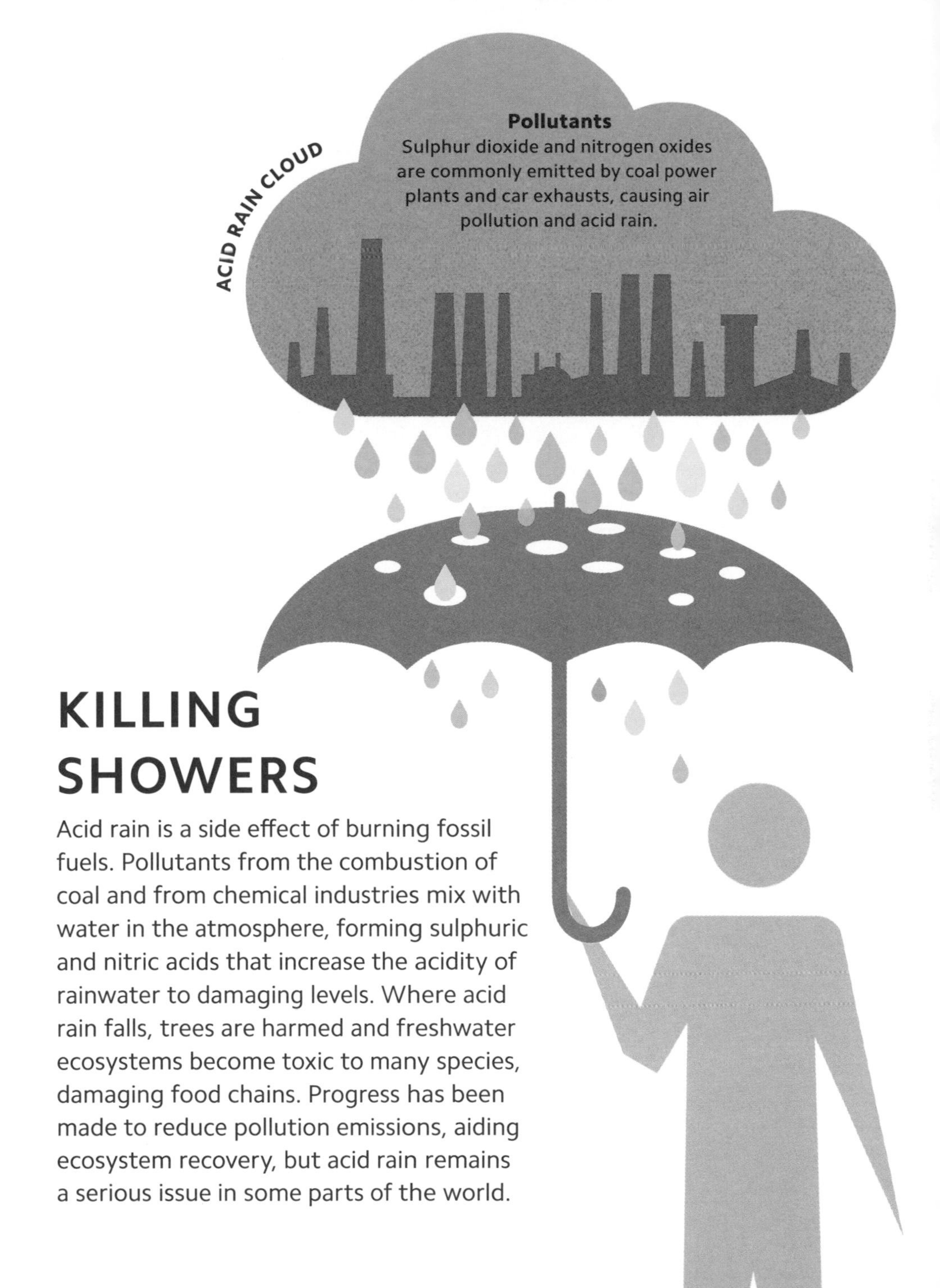

ACID RAIN CLOUD
Pollutants
Sulphur dioxide and nitrogen oxides are commonly emitted by coal power plants and car exhausts, causing air pollution and acid rain.

KILLING SHOWERS

Acid rain is a side effect of burning fossil fuels. Pollutants from the combustion of coal and from chemical industries mix with water in the atmosphere, forming sulphuric and nitric acids that increase the acidity of rainwater to damaging levels. Where acid rain falls, trees are harmed and freshwater ecosystems become toxic to many species, damaging food chains. Progress has been made to reduce pollution emissions, aiding ecosystem recovery, but acid rain remains a serious issue in some parts of the world.

TOO MUCH LIGHT

Since their invention in the late 19th century, electric lights have become commonplace in most parts of the world. However, the sheer number of lights in use (particularly in cities) results in light pollution, which causes many problems. As an inefficient use of energy, excessive use of electric lighting is a serious environmental issue. However, light pollution also has ramifications for human lifestyles and ecological systems, with negative impacts for both humans and animals.

Human impact
Light pollution is known to upset human sleep patterns by disturbing the natural rhythms of the body, making it difficult to sleep.

83 percent of the global population live in areas with light-polluted night skies.

Energy waste
Across homes, businesses, and communities, leaving the lights on is a huge cause of energy waste—in empty rooms, in displays in closed stores, and in overlit public areas.

Disturbed wildlife
Light pollution can confuse animals, disrupting their daily cycles of activity. For example, sea turtle hatchlings using the natural horizon to find the sea can be disorientated by lights.

TOXIC AIR

In addition to emitting greenhouse gases, burning fossil fuels also produces physical pollutants. Coal power plants and gasoline-fueled vehicle engines are the greatest sources of pollution. Pollution is comprised of physical particles smaller than 10 microns in diameter, which are small enough to be inhaled into the lungs. These particles cause respiratory illnesses, which are estimated to contribute to over 8 million deaths a year worldwide.

DANGER ZONES

About 91 percent of the world's population live in areas that have air pollution levels greater than the World Health Organization's guidelines.

CAN CAUSE BRAIN DAMAGE

Invisible killer

Pollution contributes to a range of health conditions, particularly heart and lung diseases. Eliminating fossil fuels will have a direct health benefit.

HEART AND LUNG PROBLEMS

MAJOR CAUSES

Regions with high densities of cars and coal-fired power stations have the greatest levels of air pollution.

HOLE IN THE SKY

Discovered in 1985 across parts of Antarctica, a hole in the ozone layer in the stratosphere (see p.17) confirmed that stratospheric ozone depletion was occurring around the world. Ozone plays an important role by blocking ultraviolet radiation, originating from the sun, from Earth. Ozone depletion was caused by the human emission of chemical particles called chlorofluorocarbons (CFCs), used in refrigerators and aerosol cans.

OZONE PRODUCTION

Ozone is produced when oxygen molecules absorb UV radiation from the sun. CFCs and other chemicals disrupt this reaction, preventing ozone from being created.

OZONE LAYER

Shrinking

The ozone hole is shrinking and the use of harmful CFCs have been banned, but it may take until 2070 for the ozone layer to recover fully.

EARTH

AT THE SURFACE

Stratospheric ozone is essential for protecting people against ultraviolet radiation, which can cause skin cancer and sunburn.

E F F E

O N L A

C T S

N D

The effects of rising temperatures are already leaving visible scars on Earth's surface. Favorable conditions for wildfires are more common, and shifting rainfall is causing droughts and desertification. This rapid change in climatic conditions makes life difficult for humans and animals, whose habitats are simultaneously being removed by human expansion. Warming is causing the ice sheets of Greenland and Antarctica to lose mass and glaciers to collapse into the ocean, contributing to a rise in the sea level. Arctic soils, which have been frozen for hundreds of years, are thawing, potentially releasing greenhouse gases such a methane, exacerbating warming.

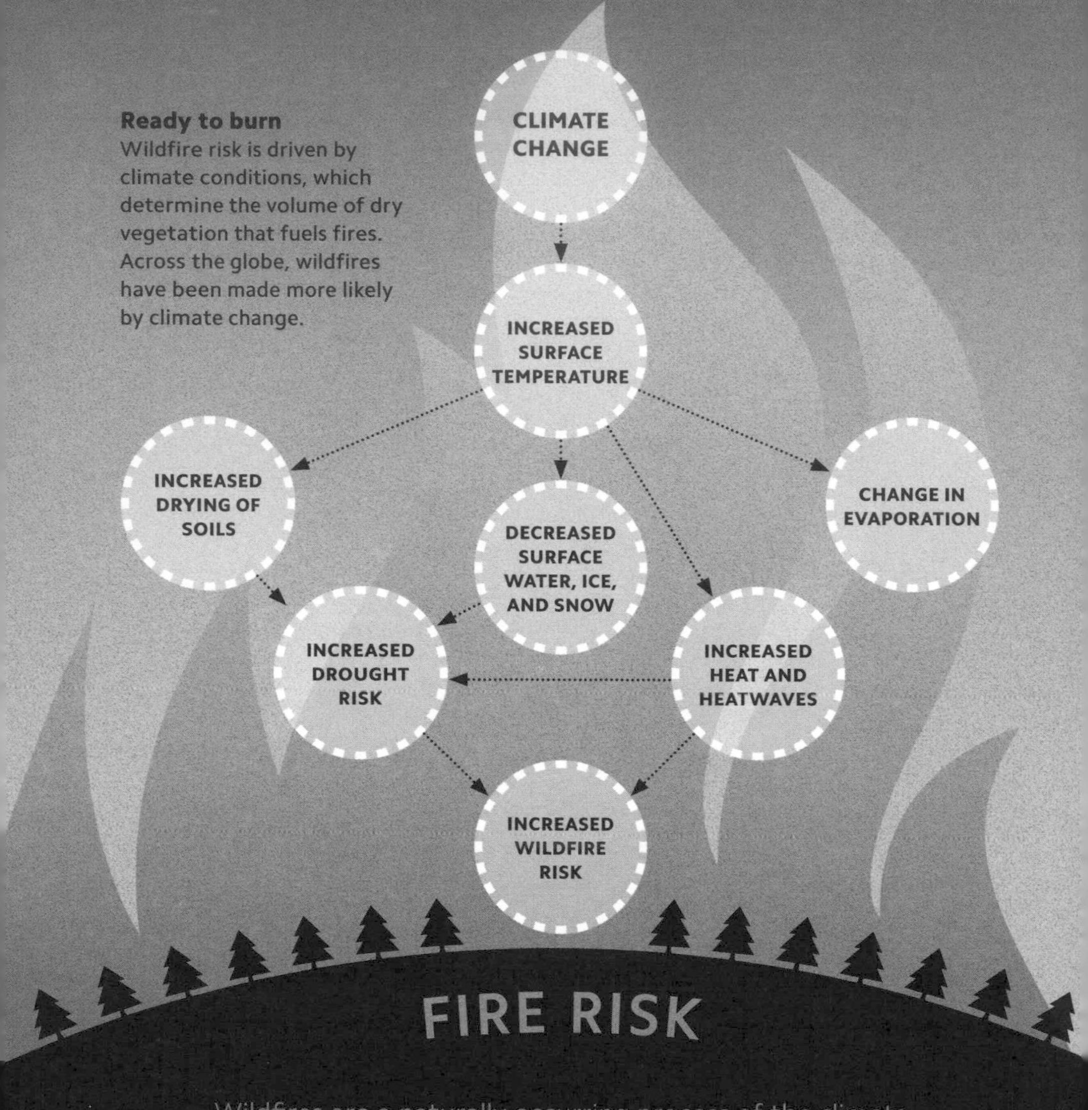

Ready to burn
Wildfire risk is driven by climate conditions, which determine the volume of dry vegetation that fuels fires. Across the globe, wildfires have been made more likely by climate change.

FIRE RISK

Wildfires are a naturally occurring process of the climate system that have been amplified by global warming. Warmer temperatures dry the soil and vegetation faster, creating ideal conditions for fires. Reduced rainfall in arid areas further enhances fire-risk conditions. Longer, more intense fire seasons across the world release huge amounts of CO_2, creating a positive feedback (see p.21), reinforcing warming.

TURNING TO DUST

A major concern for warm regions is the spread of desertification, which refers to the degradation of drylands. Soils in these arid regions are highly sensitive to shifts in moisture. Climate change causes disruption in rainfall patterns, typically making dry areas drier, and increases the frequency of drought events. This has led to the expansion of areas with soil that is unable to sustain vegetation, impacting livelihoods and food supplies.

3 billion people live in drylands across the world.

Drylands
Dryland areas already face water scarcity, and their desertification is reducing agricultural yields, with knock-on economic impacts in some of the world's most deprived regions.

FACING EXTINCTION

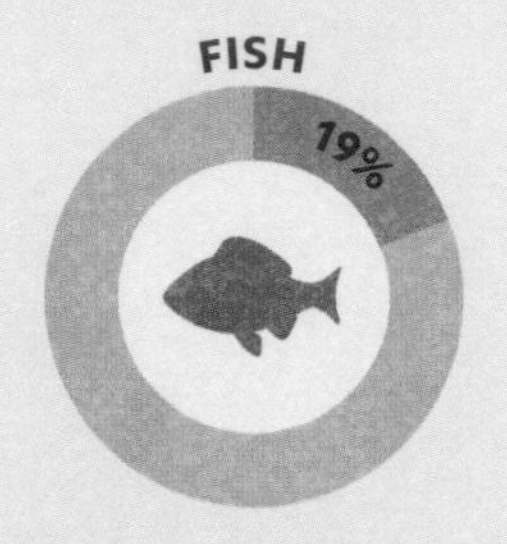

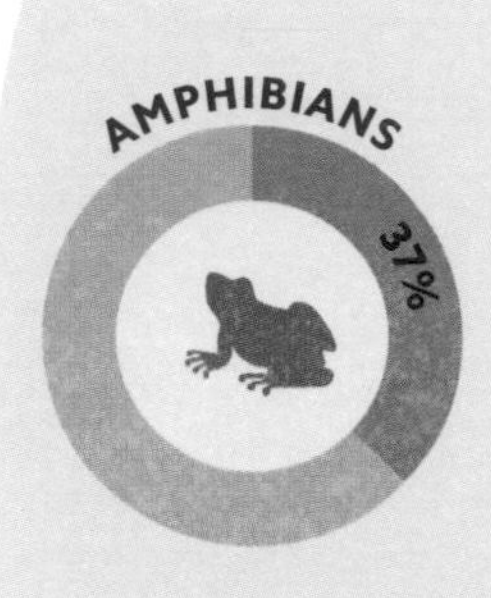

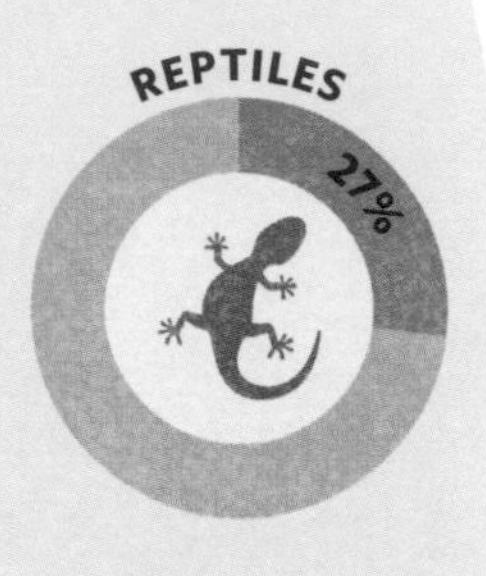

Biodiversity is the variety of life in all its forms. Human activity, such as land-use change and pollution, is causing ecosystems to degrade and shrink, reducing the number and variety of species. The effects of climate change are adding pressure to this crisis, and many plant and animal species have become extinct or shifted their habitats, where they then compete with established species.

BIRDS
23%

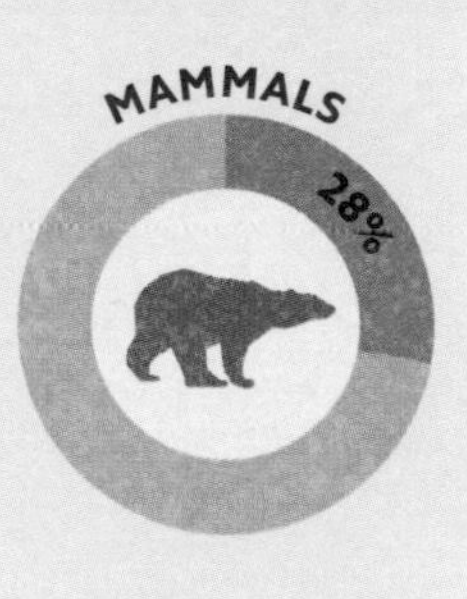

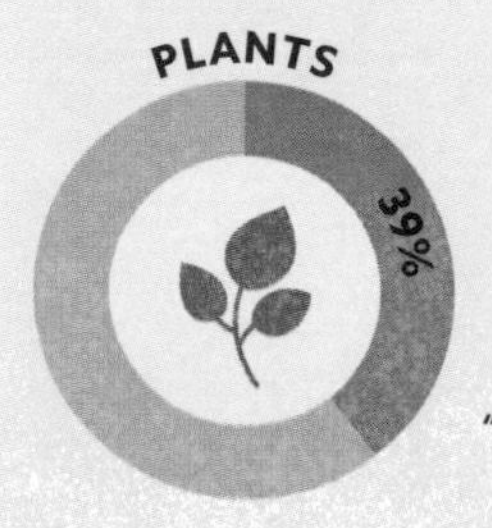

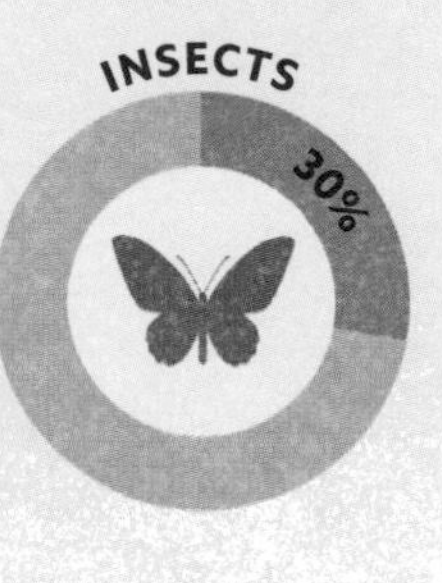

Species under threat
Here is the percentage of evaluated species in each category classified as "threatened" by the International Union for Conservation of Nature (IUCN).

NO PLACE TO CALL HOME

The extinction of animals and plants around the world is driven primarily by habitat destruction. As developers tear down natural forest habitats to make way for agricultural land and cities, the space for diverse ecosystems shrinks. Many species live in specific habitat niches, which are largely determined by climate factors such as temperature and precipitation. With climate change comes rapid shifts in the global distribution of these niches, forcing species across the world to migrate, adapt, or die out.

FLEEING DESTRUCTION

Across the world, species are being forced to migrate to smaller patches of their habitats.

Land clearances

Human activity, primarily deforestation, is a huge contributor to the destruction of habitats.

MELTING ICE

In the polar regions, the enormous ice sheets of Greenland and Antarctica are feeling the effects of a warming world. Glaciers, which flow from the interior to the coastal edges and out over the water to form ice shelves, are retreating at record rates. In the process they shed huge volumes of ice into the oceans, contributing to the rise in sea levels (see p.96). Future mass loss from Antarctica (see p.89), where marine-terminating glaciers are affected by warming ocean currents, is expected to be a major contributor to future sea-level increase.

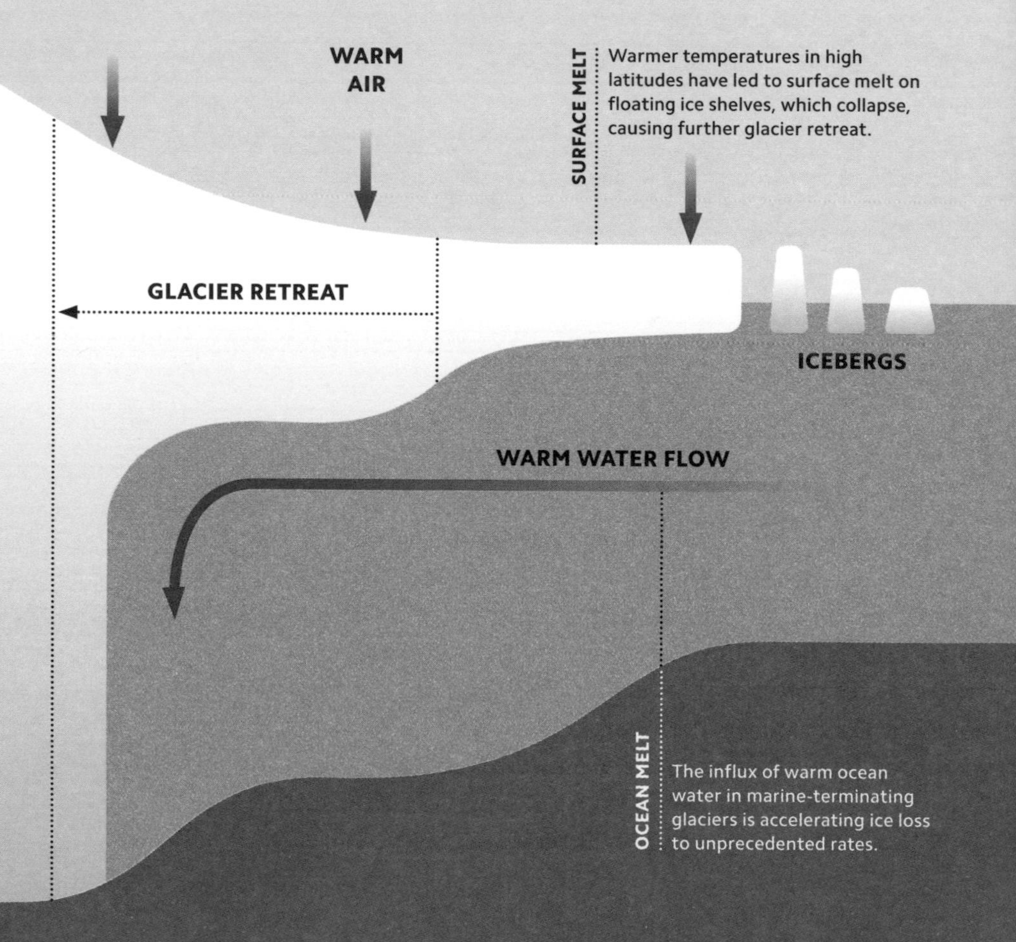

WARM
AIR
SURFACE MELT
Warmer temperatures in high latitudes have led to surface melt on floating ice shelves, which collapse, causing further glacier retreat.
GLACIER RETREAT
ICEBERGS
WARM WATER FLOW
OCEAN MELT
The influx of warm ocean water in marine-terminating glaciers is accelerating ice loss to unprecedented rates.

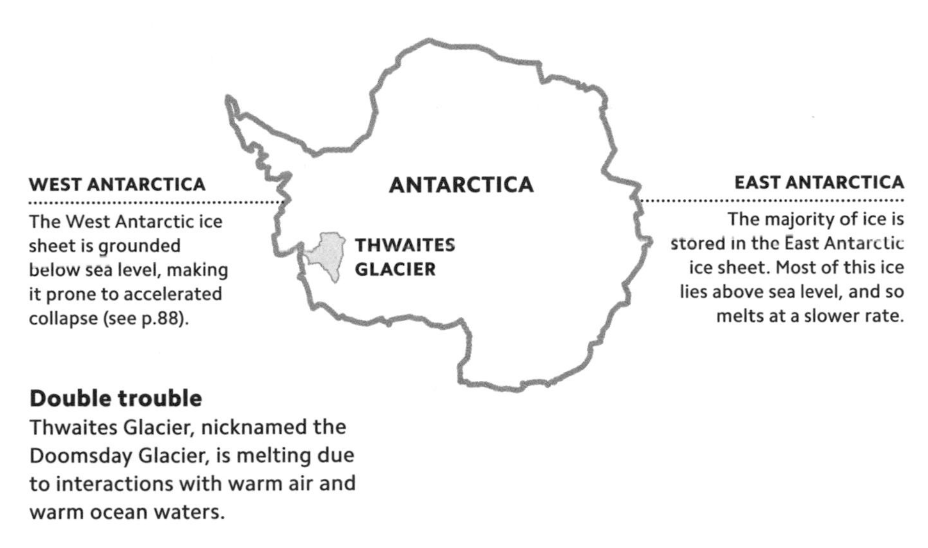

WEST ANTARCTICA
The West Antarctic ice sheet is grounded below sea level, making it prone to accelerated collapse (see p.88).
ANTARCTICA
THWAITES GLACIER
EAST ANTARCTICA
The majority of ice is stored in the East Antarctic ice sheet. Most of this ice lies above sea level, and so melts at a slower rate.

Double trouble
Thwaites Glacier, nicknamed the Doomsday Glacier, is melting due to interactions with warm air and warm ocean waters.

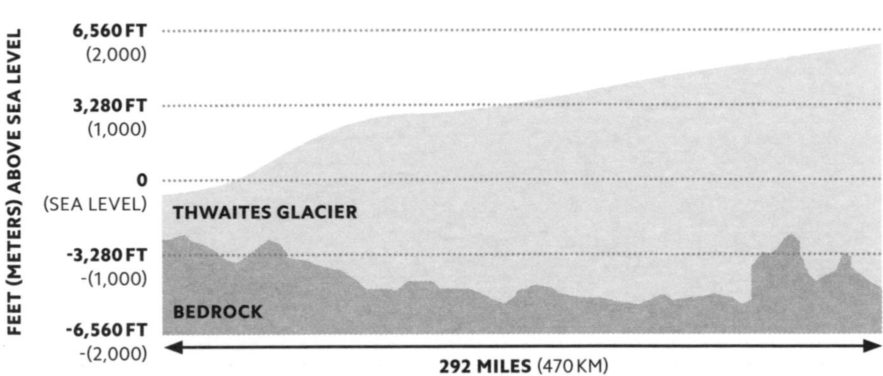

THE DOOMSDAY GLACIER

The Antarctic ice sheet holds enough ice to raise global sea levels by 90 ft (58 m) if it all melted. While this is improbable, many of its glaciers are losing ice at an accelerating rate. Of particular concern is the enormous Thwaites Glacier in West Antarctica. The ice melt of Thwaites is already responsible for more than 4 percent of global sea rise each year. If it were to melt entirely, global sea levels would rise by 2.1 ft (0.65 m).

AT MELTING POINT

Every summer, the Greenland ice sheet experiences ice melt, and every winter, fallen snow compacts and forms new ice. This seasonal melt cycle is normal, but satellite measurements reveal that the ice sheet is losing more ice in the summer melt season than it can replenish in the winter. The Greenland ice sheet contains enough frozen fresh water that if it melted, it would raise global sea levels by more than 23 ft (7 m).

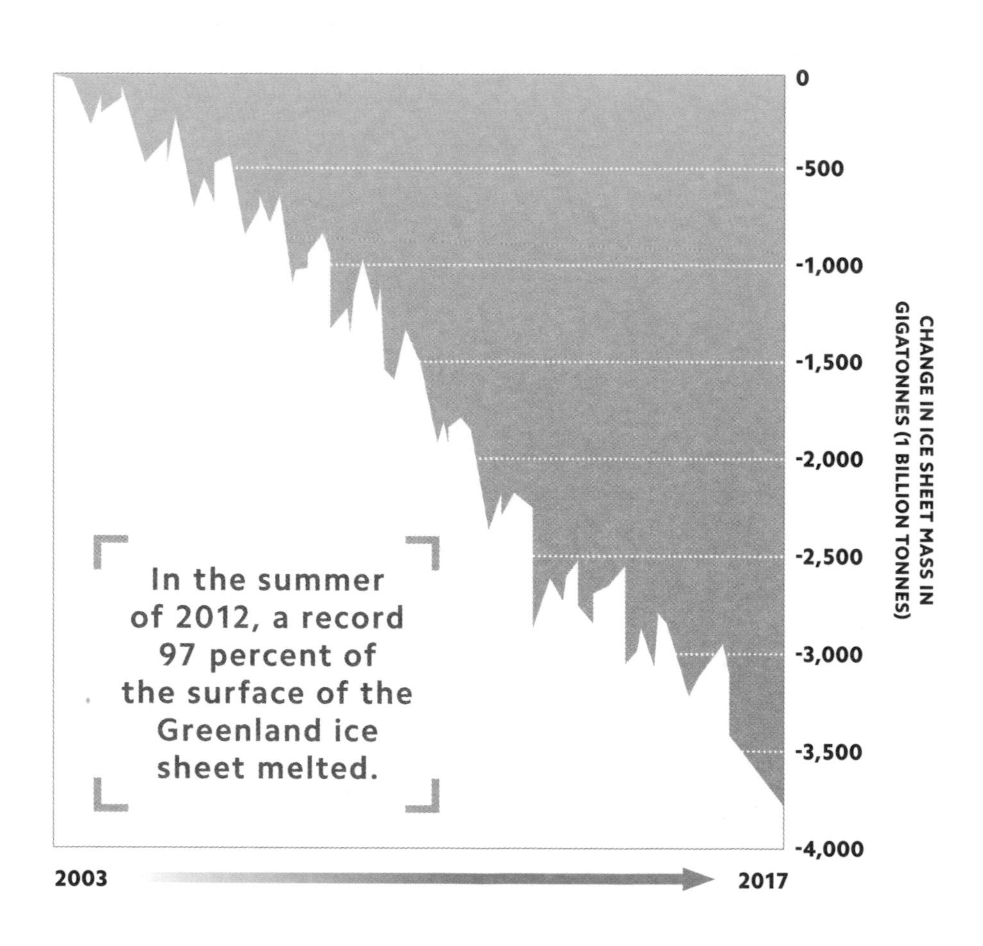

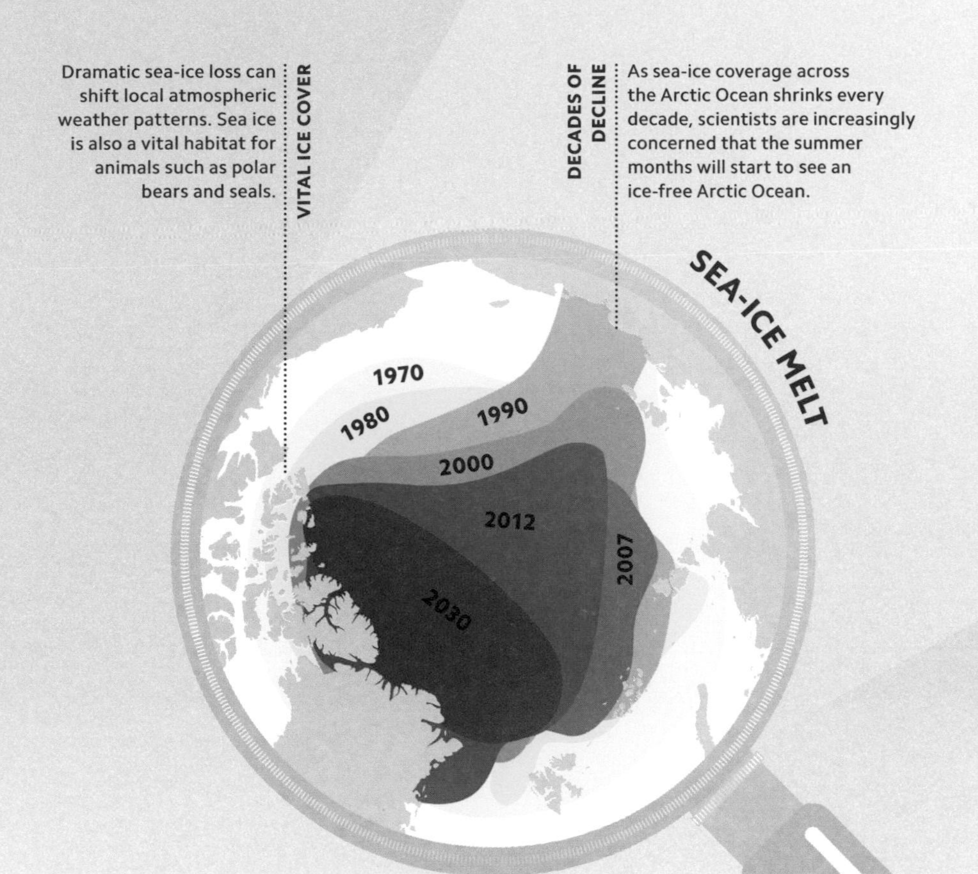

VITAL ICE COVER
Dramatic sea-ice loss can shift local atmospheric weather patterns. Sea ice is also a vital habitat for animals such as polar bears and seals.
DECADES OF DECLINE
As sea-ice coverage across the Arctic Ocean shrinks every decade, scientists are increasingly concerned that the summer months will start to see an ice-free Arctic Ocean.
SEA-ICE MELT
1970
1980
1990
2000
2012
2007
2030

ARCTIC SUMMER

Arctic sea ice is produced by the freezing of open ocean waters. The extent and thickness of the ice has been decreasing every decade since satellite records began. Sea ice reflects sunlight and as sea ice declines, the darker ocean surface absorbs more energy, which has a warming effect and closes a feedback loop, where the increase in temperature is amplified and leads to more ice melting. The result of this has been that the Arctic has warmed more than any other region on Earth.

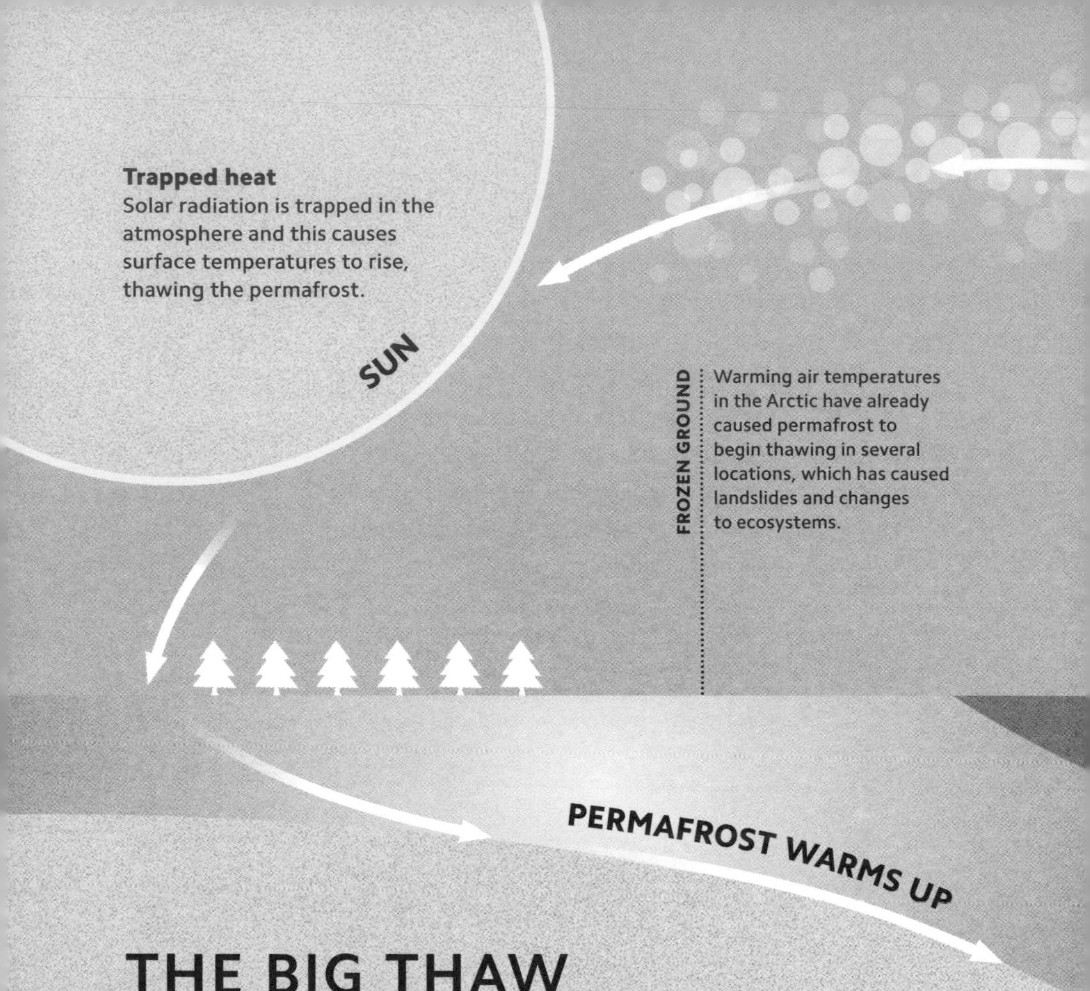

Trapped heat
Solar radiation is trapped in the atmosphere and this causes surface temperatures to rise, thawing the permafrost.

SUN

FROZEN GROUND

Warming air temperatures in the Arctic have already caused permafrost to begin thawing in several locations, which has caused landslides and changes to ecosystems.

PERMAFROST WARMS UP

THE BIG THAW

Permafrost refers to soil that remains frozen for two or more consecutive years. The majority of global permafrost is in Scandinavia, Siberia, Alaska, and the northern parts of Canada, regions that are heating at twice the global average. As permafrost thaws, the abundance of organic matter trapped within it is decomposed by microbes, which releases greenhouse gases into the atmosphere. Scientists estimate that there are 1,400 gigatons of carbon trapped in permafrost, almost twice as much as there is currently in the atmosphere.

FEEDBACK LOOP

Carbon and methane emissions increase the greenhouse effect, causing warming, and further permafrost thawing.

CARBON AND METHANE

Carbon dioxide (CO_2) and methane (CH_4) permeate the soil, seep through the water, and are released into the atmosphere.

CH_4

CO_2

WATER

EMISSIONS

GREENHOUSE GAS RELEASE

Organic matter is digested by microbes, producing methane and carbon dioxide.

DECOMPOSITION

Below the surface, organic matter and previously frozen microbes thaw.

EFFEC
THE **OC**

T S O N
E A N S

The world's oceans are under assault from physical and chemical changes. While the main cause of ocean heat is sunlight, clouds, water vapor, and greenhouse gases in the atmosphere also emit heat, some of which is absorbed by ocean water. This contributes to marine heatwaves and the thermal expansion of the ocean, leading to a rise in sea level. Marine ecosystems, such as coral reefs, are threatened by high ocean temperatures and some fish stocks have already moved toward the poles. Acidification of the oceans, caused by carbon dioxide (CO_2) dissolving in the water, is also damaging marine organisms.

SURGING SEAS

Since tide gauge measurements began, the average sea level has been rising. Recent improvements in scientific understanding have attributed global sea-level rise to two main factors: ice melt and thermal expansion. As ocean water absorbs thermal energy trapped by the greenhouse effect, it warms and increases in volume. In addition, there has been a large net transfer of water stored on land, in the form of ice, into the ocean.

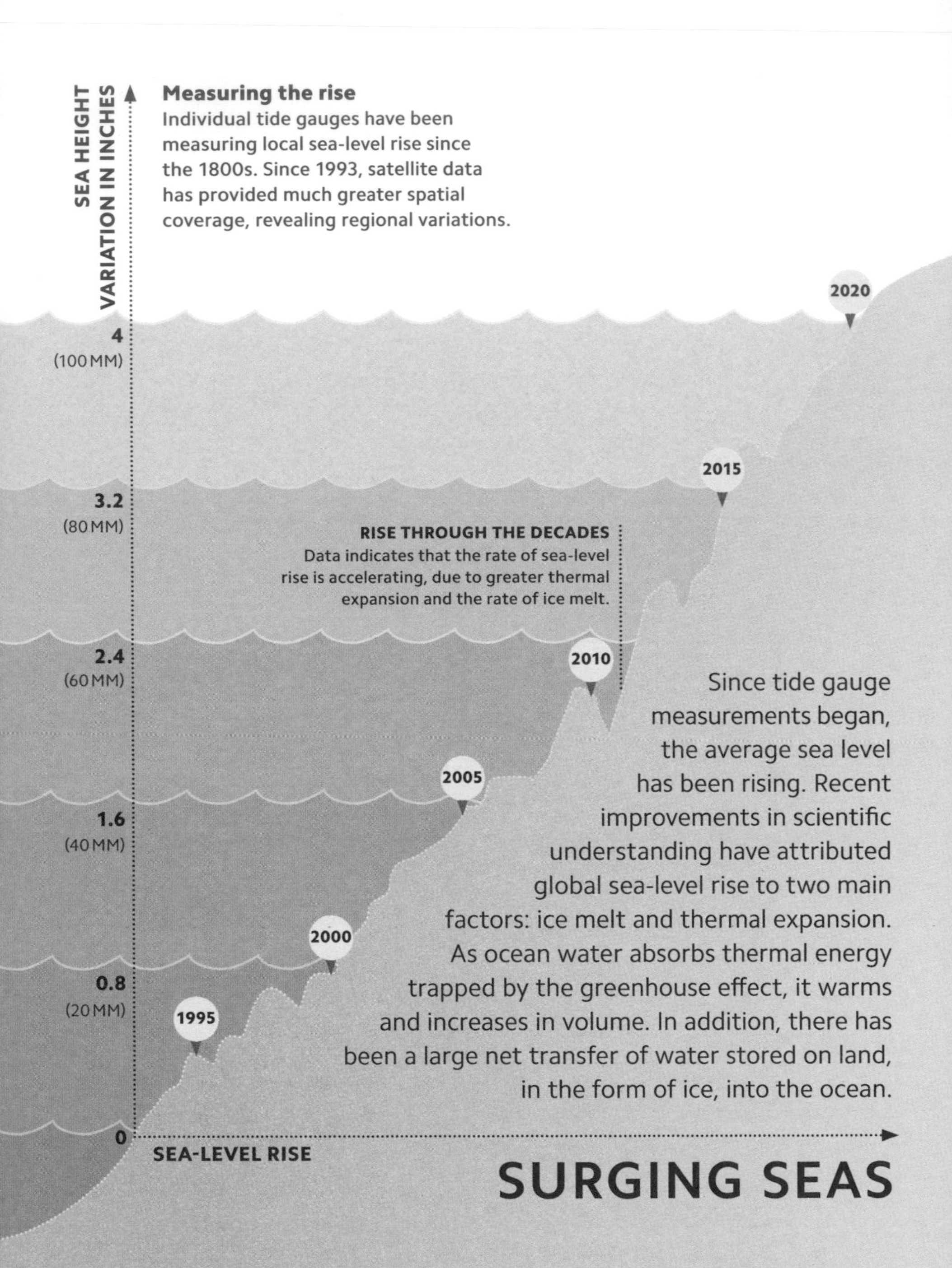
SEA HEIGHT VARIATION IN INCHES
Measuring the rise
Individual tide gauges have been measuring local sea-level rise since the 1800s. Since 1993, satellite data has provided much greater spatial coverage, revealing regional variations.
4 (100 MM)
3.2 (80 MM)
2.4 (60 MM)
1.6 (40 MM)
0.8 (20 MM)
0
RISE THROUGH THE DECADES
Data indicates that the rate of sea-level rise is accelerating, due to greater thermal expansion and the rate of ice melt.
1995
2000
2005
2010
2015
2020
SEA-LEVEL RISE

DROWNING ISLANDS

16.5 FT (5 M)

Nowhere will the impact of the current rate of sea-level rise be more devastating than on low-lying islands. These nations have contributed very little in terms of emissions, but are facing the most severe consequences, as the land on which they live is in danger of disappearing. Nations such as the Maldives in the Indian Ocean and Kiribati in the Pacific contain many islands that do not rise more than 3.3 ft (1 m) above sea level.

156.6 MILLION DISPLACED

Coral islands

Low-lying islands are typically founded on coral reefs. The constant movement of sediment by ocean currents means the precise shape of each island is constantly changing. Scientists hope natural processes will keep raising these islands as the sea rises.

3.3 FT (1 M)

IMMINENT DANGER

Governments of threatened islands are already making plans to relocate populations to larger, neighboring islands, as the rate of sea-level rise is increasing.

15 MILLION DISPLACED

SEA LEVEL

WHEN ALL THE ICE HAS MELTED

While the predicted height of sea-level rise may seem trivial (see pp.96–97), the number of people in coastal communities it would affect is huge. Although sea level is not predicted to rise more than 3.3 ft (1 m) by 2100, scientists believe that the last time CO_2 concentration was at 2020 levels, sea level was 66 ft (20 m) higher than it is now. As waters rise, the likelihood of dangerous floods in major urban areas increases. Storm surges are already causing deadly flooding events, and these will become stronger and more frequent as sea level rise increases.

200 million people currently live in homes that could be below the high-tide line in 2100.

Cities at risk
Many of the world's major cities are situated in coastal areas, and a significant rise in sea level could bring devastation to densely populated urban centers.

230 FT (70 M)

LOST LAND
Large parts of the coast could be flooded permanently, or be impacted negatively by advancing sea water.
AMAZON BASIN
A rise of 230 ft (70 m) could see the Amazon Basin becoming an inlet of the Atlantic Ocean, destroying vast areas of rainforest.
LIMA
SOUTH AMERICA
RIO DE JANEIRO
SUBMERGED CITIES
A rise of more than 215 ft (66 m) could cause cities on the River Plate estuary to disappear under water.
BUENOS AIRES
MONTEVIDEO

Shifting coasts
If all the world's ice melted, sea level could rise by as much as 230 ft (70 m), dramatically altering the coastlines of many countries. However, such a rise would happen over a very long time period, giving vulnerable nations time to adapt.

GREENHOUSE EFFECT

Greenhouse gases trap heat, which is absorbed by both oceans (blue) and atmosphere (green).

TRAPPED GASES

SEA CHANGE

The oceans are bearing the brunt of human-induced climate change. The vast majority of the excess heat trapped through the greenhouse effect (see p.12) is absorbed by the ocean as energy. This energy is absorbed at the oceans' surfaces before being redistributed by deep currents that circulate around the world. This means that heat trapped by greenhouse gases has reached the deepest oceans and the coldest Antarctic currents.

OCEANS ABSORB 90% OF EXCESS HEAT FROM GREENHOUSE GASES

Taking the temperature
Measuring the amount of heat stored in the oceans is a technical challenge, with measurements mainly taken by specifically designed drifting buoys that sink and rise to record temperatures at varying depths.

ATMOSPHERE ABSORBS HEAT

Only 10 percent of excess heat from the greenhouse effect is absorbed by the atmosphere. This causes changes in air temperature.

DEADLY OIL SPILL

This oil spill in the Gulf of Alaska was one of the largest oil spills in history. It devastated local wildlife, and prompted 11,000 local residents to offer their help in cleanup efforts.

DANGEROUS HEAT

One of the most extreme impacts of rising ocean temperatures is marine heatwaves. These occur when ocean temperature is above the typical seasonal range for a prolonged period—usually five consecutive days. Extreme events such as marine heatwaves put pressure on marine ecosystems, and the animal and human communities that are dependent on them. Since the advent of human-caused global warming, the likelihood of large marine heatwaves in susceptible areas has increased by 20 times.

KILLER HEATWAVE

In the Gulf of Alaska, a record-breaking heatwave reduced the number and quality of phytoplankton. This disrupted the food web, causing many marine organisms, such as seabirds, to die off in great numbers. This ecosystem damage exceeded that of the devastating oil spill years before.

ELEVATED CO_2 LEVELS

0.1 IN (3 MM) SEA LEVEL RISE A YEAR

1–7% DECLINE IN OXYGEN CONTENT BY 2100

REDUCED OCEAN MIXING POSES ADAPTION CHALLENGES FOR MARINE LIFE

50% POTENTIAL DECLINE IN ANNUAL CATCHES

Marine malfunction
The effect of climate change is drastically changing the state of the oceans, making life much harder for many species and also for people whose livelihoods depend on the oceans.

ENDANGERED ECOSYSTEM

Marine species affected by the impact of climate change include plankton—the basis of marine food chains—corals, fish, polar bears, walruses, seals, sea lions, penguins, and other seabirds. A decline in one species has implications for the rest of the ecosystem, and the pressure from a number of climate change stresses is predicted to increase. Climate change could therefore cause the extinction of many species already under stress from overfishing and habitat loss.

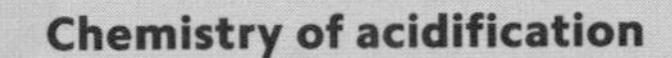

Chemistry of acidification

As more CO_2 dissolves in the ocean, it combines with water to form carbonic acid (H_2CO_3-). This separates, forming ions of hydrogen and bicarbonate, which increase acidity.

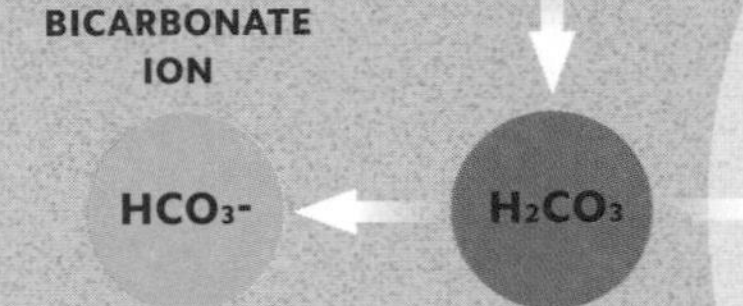

Seawater reaction

The absorption of CO_2 in the oceans has raised the surface water's acidity by 30 percent.

DISTURBING CHEMISTRY

In addition to energy, the ocean absorbs carbon dioxide (CO_2) directly from the atmosphere. As CO_2 is added to the atmosphere, the ocean absorbs approximately 30 percent of it. This increased CO_2 content in the ocean reduces the water's pH value, driving ocean acidification. Many marine organisms, such as corals and mollusks, build their shells using carbonate ions, which are reduced in concentration by ocean acidification. Only a small number of organisms, such as seagrasses, may benefit from the increased acidity.

REEFS UNDER THREAT

Caused by excessively warm ocean temperatures (see p.100), coral bleaching occurs when stressed corals expel the colorful algae that live inside them. Coral can survive a bleaching event, but if the heat stress is prolonged, the coral dies, and the ecosystem dependent on it often collapses. Fatal coral bleaching is one of the most immediate dangers of climate change, with 50 percent of the Great Barrier Reef having died since 2016.

BLEACHED WHITE When exposed to heat stress, coral ejects its algae (zooxanthellae), causing the coral to turn a pure white. At this stage, the coral can still recover.

DEAD AND DECAYING Without algae, the coral starves. If this is prolonged, the coral dies, and eventually decays. As a result, coral reef organisms permanently lose their habitat.

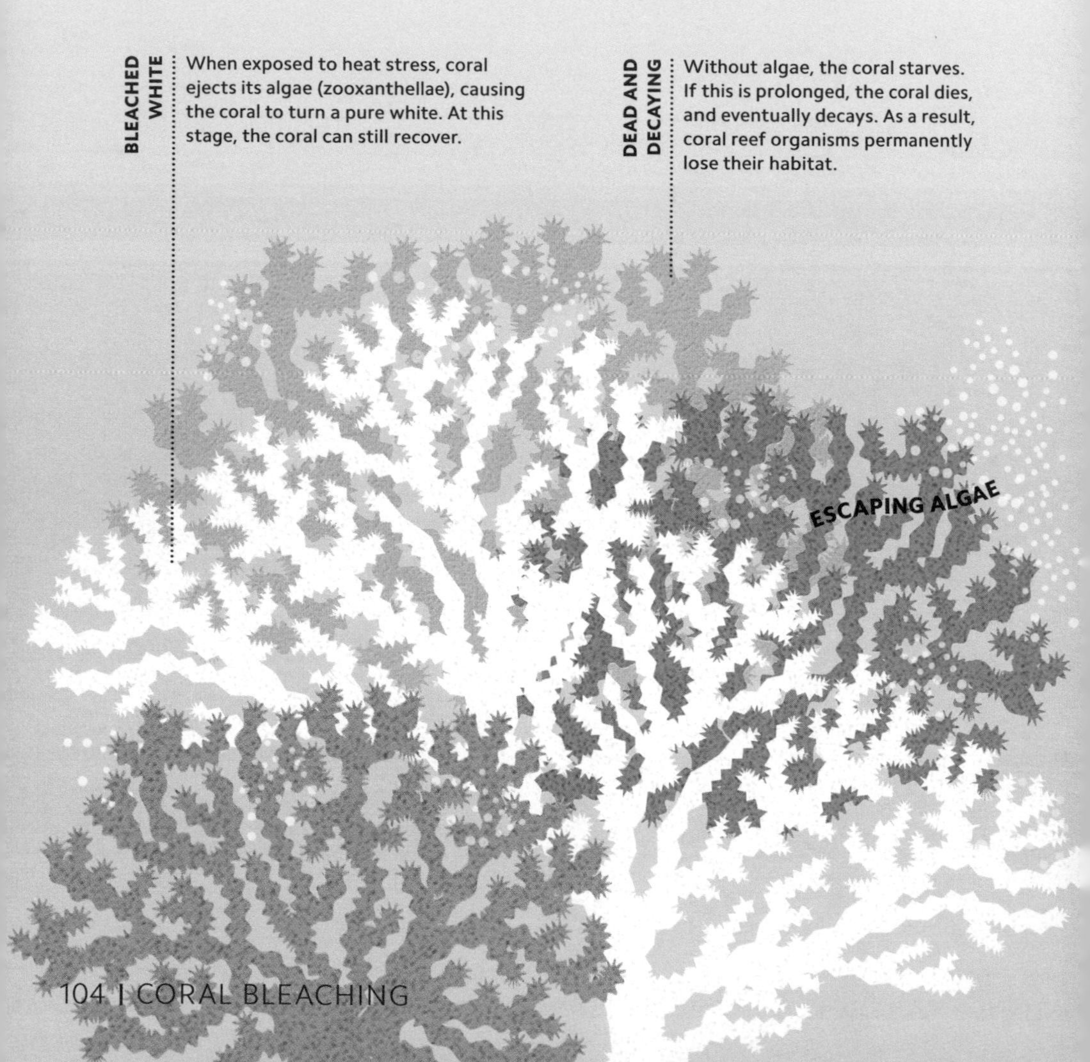

Pouring in

More than 419 million tons (380 million tonnes) of plastic is produced annually. Of this, up to 14.3 million tons (13 million tonnes) enters our oceans every year.

DISCARDED PLASTIC

PLASTIC WASTE

It is estimated that 80 percent of plastics in the ocean were used on land. The rest is thought to come from ships at sea.

HARMING ANIMALS

Ocean plastic does huge amounts of damage to the marine ecosystem, as animals can ingest the plastic or become entangled and suffocate.

Plastic pollution in the oceans has been a growing problem since the 1950s. A huge volume of plastics has found its way into our oceans, and as a nonbiodegradable material, it remains in circulation indefinitely. Most plastics reach the ocean as a result of inefficient disposal methods. These methods lead to plastic items of all sizes ending up in the ocean, endangering marine wildlife.

FOOD CHAIN

PLASTIC SOUP

HUMA
COST

N

The impacts of climate change are not spread evenly. Everywhere, the fragile relationships between humans and the land they farm and live on face rising challenges from climate change. However, the populations facing the greatest threats are typically those in less wealthy nations, who have contributed the least to greenhouse gas emissions. As climate change worsens natural disasters, populations are expected to migrate away from hotspots, creating political challenges. Food and freshwater security are the most immediate threats, as agricultural systems are crippled by drought and extreme temperatures and the ocean's ecosystems collapse.

THE HARDEST HIT

The impacts of climate change are not evenly distributed. In many instances, people living in nations that have contributed the least emissions historically are those facing the most severe impacts of climate change. For example, many of the nations most vulnerable to rising seas, stronger storms, and deadly heatwaves are small or island nations in the tropics. In contrast, populations with wealthy, high-emitting lifestyles are often less susceptible to the immediate effects of climate change.

PUERTO RICO

MOZAMBIQUE

THAILAND

PHILLIPINES

THE BAHAMAS

The least protected
These ten lower-income regions were rated the worst-affected by extreme weather events from 2000–2019. Due to costs, they often cannot enact the same large-scale protection measures available to higher-income nations.

MYANMAR

PAKISTAN

NEPAL

HAITI

BANGLADESH

SUB-SAHARAN AFRICA

86 MILLION

CENTRAL AND SOUTH AMERICA

17 MILLION

SOUTH ASIA

40 MILLION

Forced from their homes
If climate change continues unabated, then by 2050 there could be more than 143 million internal climate migrants in total within these three regions.

DISPLACED BY DISASTERS

As climate stresses make human life more difficult, migration away from heavily afflicted and often agriculture-dependent areas becomes a likely consequence of climate change. While attributing a firm cause to the redistribution of human populations within a region is difficult, living conditions are expected to gradually worsen in regions suffering heatwaves and in coastal areas susceptible to sea-level rise. This, alongside the rising intensity of damaging natural disasters (see p.75), is expected to cause climate migration to rise.

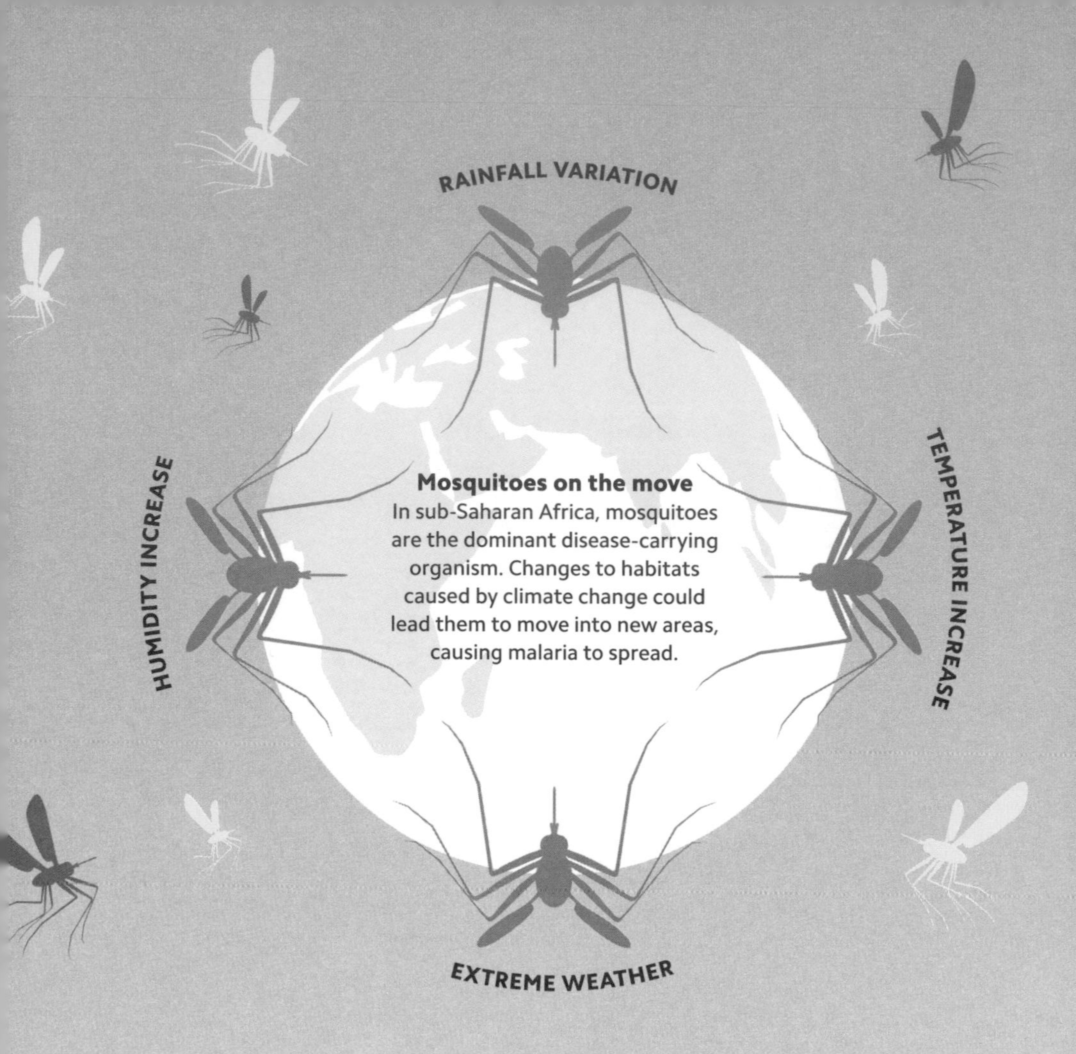

RAINFALL VARIATION
HUMIDITY INCREASE
TEMPERATURE INCREASE
EXTREME WEATHER
Mosquitoes on the move
In sub-Saharan Africa, mosquitoes are the dominant disease-carrying organism. Changes to habitats caused by climate change could lead them to move into new areas, causing malaria to spread.

THE SPREAD OF DISEASE

Surges in diseases driven by climate change are a great source of concern. One major way that previously unknown pathogens are transferred to humans is through close contact with wild animals. As climate change shifts the distribution of many animal species (see p.87), the risk of disease transmission from animal populations carrying ticks or biting insects to humans may increase. As climate change moves these organisms, it spreads the diseases they carry.

LESS FOOD IN OUR FUTURE

LESS RED MEAT

Some dietary changes, such as lower meat consumption, may result in fewer deaths.

Assessing the impact of climate change on diets across the world is important, particularly for children in the lowest-income parts of the world. A nutritious and balanced diet is key to good health, especially in the first three years of life. Crop supplies are already vulnerable to extreme weather events, which are predicted to worsen as the planet warms. Long-term variation in rainfall and temperature are set to lower crop yields and raise prices, threatening food security (see p.112).

LESS FRUITS AND VEGETABLES

A dietary shortfall in nutritious fresh fruit and vegetables is set to become a significant cause of death in the coming decades.

Climate change could increase the risk of hunger and malnutrition by up to 20 percent by 2050.

Access for all
The food system supports the livelihoods of more than one billion people, many of whom are at risk of both their food supply and economic support collapsing at the same time.

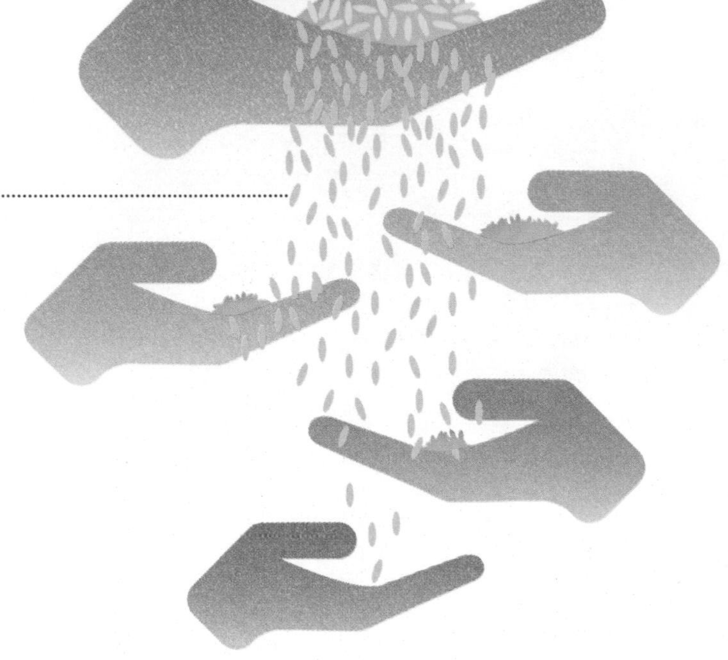

AVAILABILITY OF FOOD

Changes in a region's climate, such as a shift in the onset of spring or changes to rainfall patterns, may affect the region's ability to grow its own food, making it reliant on being able to buy food from other nations.

THE STRUGGLE FOR FOOD

Food security is a measure of the availability of sufficient and nutritious food in a country or region. Food insecurity is one of the most direct consequences of climate change. The Intergovernmental Panel on Climate Change (IPCC) believes that climate change is already weakening food security in several regions. Crop regions around the equator have been especially vulnerable to temperature and rainfall stresses, resulting in reduced crop yields and rising prices.

THIRSTY WORLD

RUNNING LOW

Rivers provide fresh water to billions of people. If rainfall patterns change, and river discharge decreases, these populations will suffer water shortages.

Freshwater resource
Much of the world's fresh water is locked up in remote glaciers and ice and is not available for consumption.

GROUNDWATER

Groundwater provides 50% of domestic water, but monitoring it is difficult, making it a challenging resource to manage.

ONLY 2.5% OF THE EARTH'S WATER IS FRESH WATER

Climate change is causing an intensification of the water cycle, meaning that floods and droughts are becoming more frequent. This will put human populations under stress of reduced water security. Droughts and desertification in drylands will continue to increase, removing the groundwater supplies. About 3 billion people are expected to be living in water-scarce regions around the world by 2050. In coastal areas, increased flooding risks increasing the contamination of vital freshwater resources.

LARGE-
SOLUTI

SCALE ONS

Fixing climate change means finding new energy sources and reducing carbon dioxide emissions to net zero (see pp.36–37). This requires large-scale changes in all sectors, from transportation to agriculture. New innovations in technology and design, shifts in public policy, and the increased use of renewable electricity are just the beginning. To achieve this requires global solutions and international cooperation. The world can also provide climate finance, so that poorer countries, which are often the hardest hit, can adapt to the changes that are already taking place today.

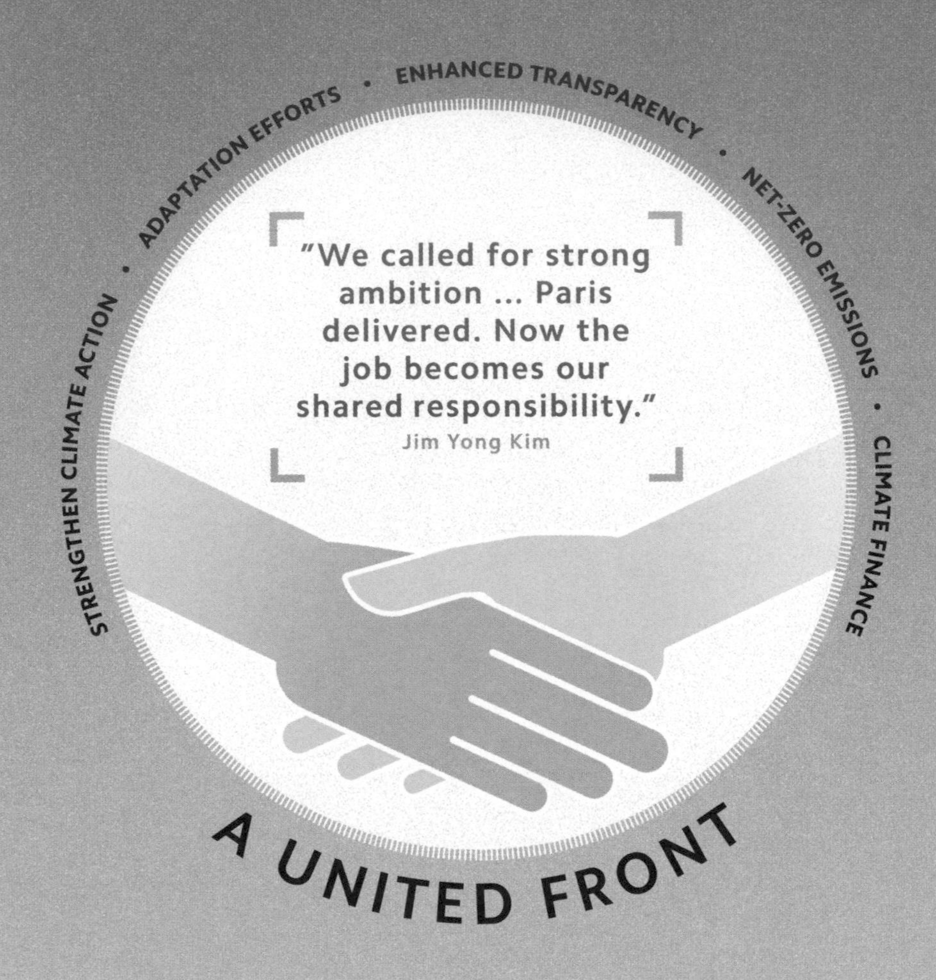

ENHANCED TRANSPARENCY
NET-ZERO EMISSIONS
CLIMATE FINANCE
STRENGTHEN CLIMATE ACTION
ADAPTATION EFFORTS
"We called for strong ambition ... Paris delivered. Now the job becomes our shared responsibility."
Jim Yong Kim
A UNITED FRONT

Averting a global climate disaster requires all countries to commit to global goals and actions, such as those of the Paris Climate Agreement (2016). This climate accord aims to limit global warming to "well below" 3.6°F (2°C). It also stresses the need for adaptation—the mitigation of current and future effects of climate change, especially for vulnerable groups—and climate finance, which involves richer nations financially supporting nations with less means of responding to climate impacts.

BALANCING THE SCALES

Climate change affects everyone, however climate change does not affect everyone equally. Poorer countries—which have emitted the least—are the hardest hit by climate change, and marginalized people are often on the front lines of environmental destruction. Climate justice aims to address these imbalances through a variety of measures. These range from taking emitters to court for climate damages, to climate finance transferring money from richer nations to support poorer nations.

POORER NATIONS

RICHER NATIONS

Bearing the brunt
Less well-off nations may suffer greater impacts and do not share equally in the income made from polluting industries.

Avoiding the effects
Countries with greater wealth may contribute more to creating climate change, but many have yet to face the same impacts as poorer nations.

LOW-CARBON GROWTH

Until recently, economic growth had been inextricably linked to rising CO_2 emissions. A recent measure of global emissions intensity shows that 27 oz (768 g) of CO_2 is emitted for every one US dollar of gross domestic product generated. Some nations exhibit considerably lower rates (for example, Japan's emissions intensity is 8.6 oz/ 244 g per 1 USD), but a much greater decoupling of emissions from growth is required globally to meet future targets while maintaining prosperity for growing populations.

GLOBAL ECONOMIC GROWTH

FALLING CO_2 EMISSIONS

Different paths
The goal of absolute decoupling sees greenhouse gas emissions fall in absolute terms, while economies continue to grow.

"There is no contradiction between sustainability and economic growth."
Valdis Dombrovskis

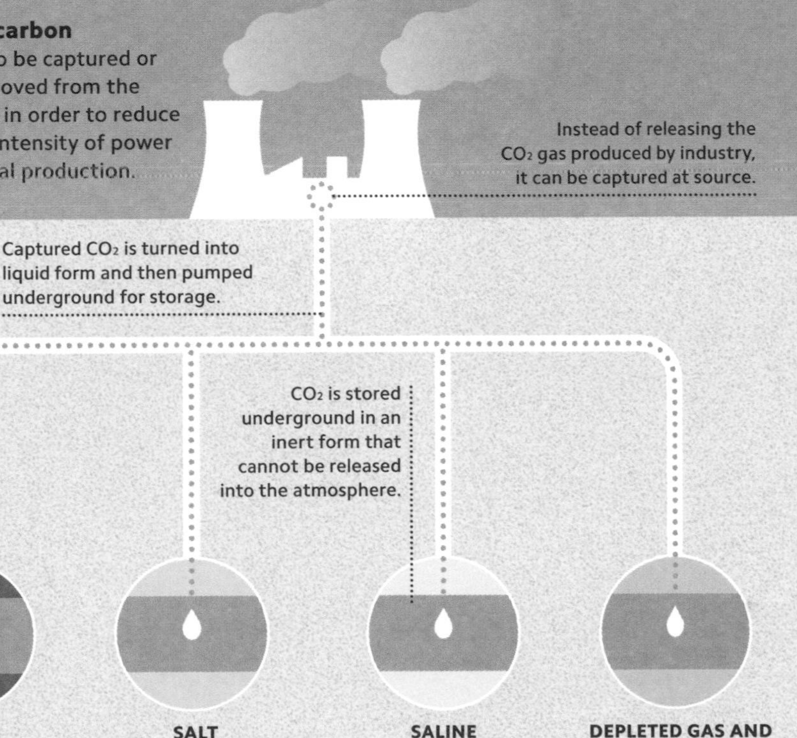

Captured carbon
CO_2 needs to be captured or actively removed from the atmosphere in order to reduce the carbon intensity of power and industrial production.
Instead of releasing the CO_2 gas produced by industry, it can be captured at source.
Captured CO_2 is turned into liquid form and then pumped underground for storage.
CO_2 is stored underground in an inert form that cannot be released into the atmosphere.
COAL BEDS
SALT BEDS
SALINE AQUIFER
DEPLETED GAS AND OIL RESERVOIRS

DEEP STORAGE

Carbon-capture technologies promise a significant reduction in atmospheric CO_2 emissions through the post-combustion capture of CO_2. The CO_2 is then liquefied, transported, and injected deep underground into suitable locations, such as saline aquifers or depleted oil reservoirs. Only a handful of carbon capture and storage (CCS) systems are currently in operation. Given the scale of present-day power and industrial production systems, it is very likely that, even with rapid decarbonization, there is likely to be a need for CCS to achieve net-zero (see pp.36–37).

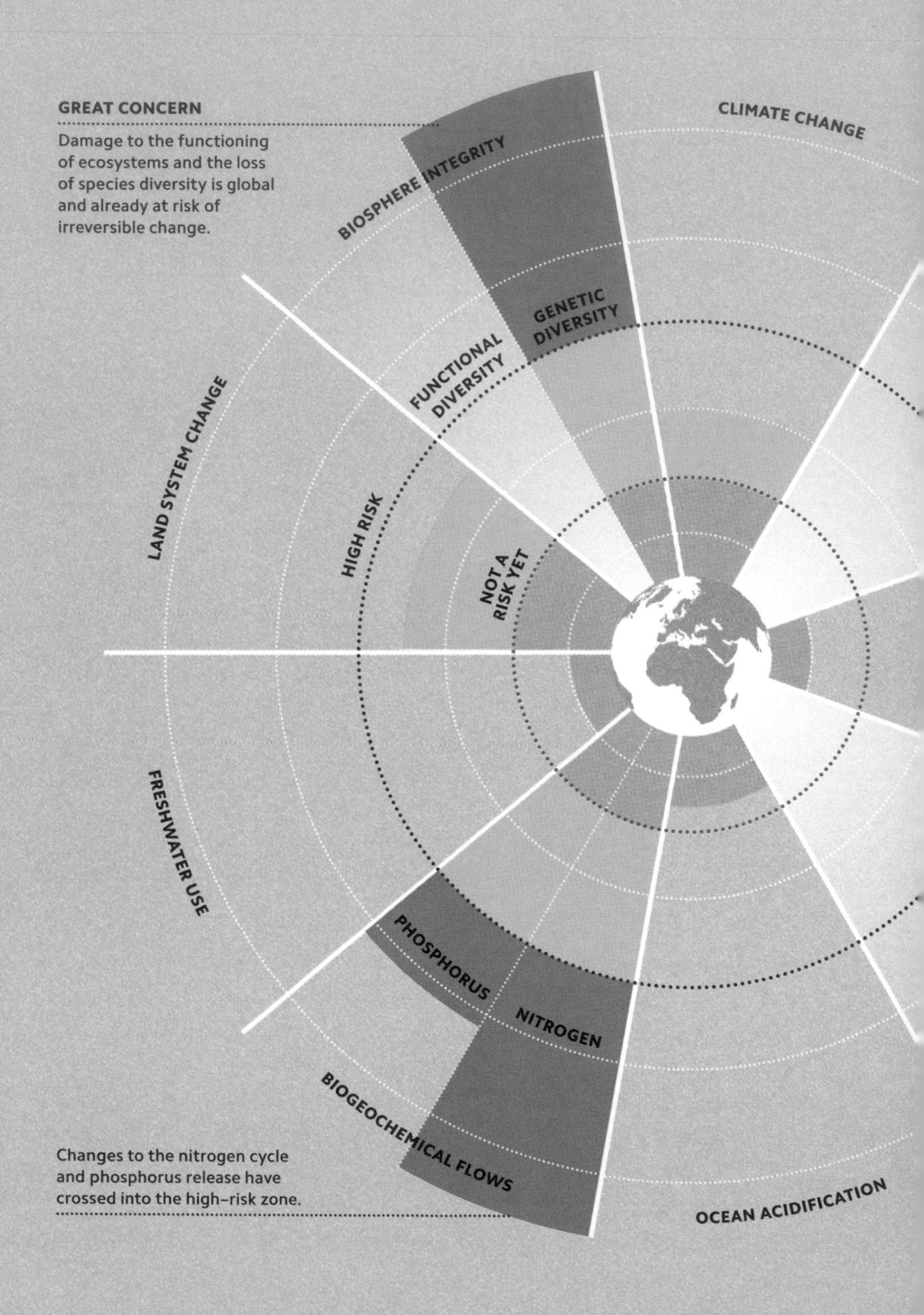
GREAT CONCERN
Damage to the functioning of ecosystems and the loss of species diversity is global and already at risk of irreversible change.
CLIMATE CHANGE
BIOSPHERE INTEGRITY
GENETIC DIVERSITY
FUNCTIONAL DIVERSITY
LAND SYSTEM CHANGE
HIGH RISK
NOT A RISK YET
FRESHWATER USE
PHOSPHORUS
NITROGEN
BIOGEOCHEMICAL FLOWS
Changes to the nitrogen cycle and phosphorus release have crossed into the high-risk zone.
OCEAN ACIDIFICATION

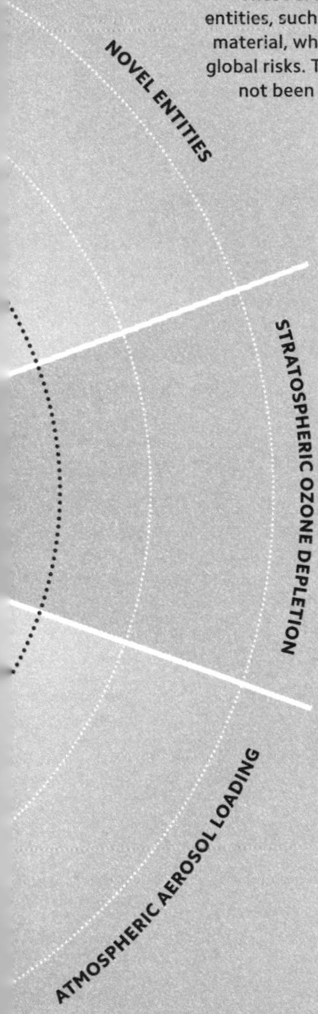

DEFINING SAFE BOUNDARIES

NOVEL RISKS

These are human-made entities, such as radioactive material, which could pose global risks. These risk have not been measured yet.

NOVEL ENTITIES

STRATOSPHERIC OZONE DEPLETION

ATMOSPHERIC AEROSOL LOADING

The Planetary Boundaries framework is a new view of the challenges facing humans and Earth that was proposed by a team of 28 scientists in 2009. Each boundary is clearly defined, such as maintaining 90 percent biodiversity or keeping atmospheric CO_2 concentrations below 350ppm (parts per million), and provides a safe space within which humanity can operate while keeping Earth systems intact. Crossing any boundary, as the current rate of 410ppm CO_2 does, increases the risk of large-scale environmental changes, many potentially irreversible.

> **"We need to now reconnect our entire world to the planet."**
> Johan Rockström

Nine boundaries
The Planetary Boundaries framework identifies nine "boundaries" and the potential hazards of crossing any of them.

CLOSING THE LOOP

The traditional linear model of production–use–disposal (also referred to as take–make–waste) is increasingly challenged by the concept of a circular economy. In the circular model, growth is not dependent on the consumption of resources that will eventually run out. It involves three core principles: ecodesign to design out waste, greenhouse gas emissions, and other pollution; keeping products, materials, and components in use and circulating in the economy through repair, reuse, recycling, and redistribution; and firm commitments to preserve, repair, and regenerate natural systems.

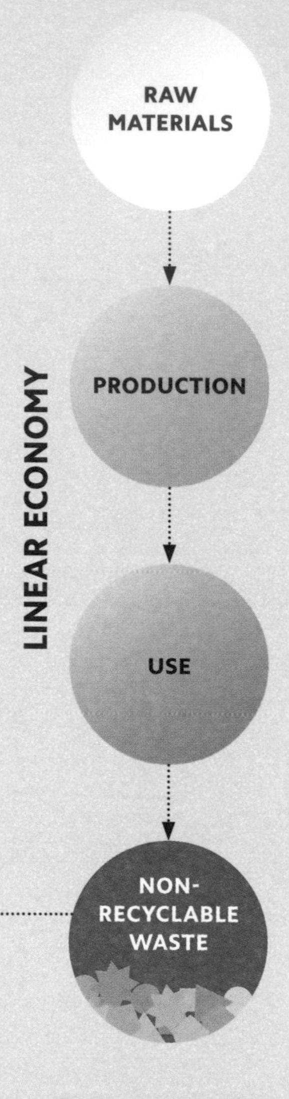

WASTE PROBLEM

The amount of waste generated by centuries of economic development has led to climate change, pollution, and widespread ecosystem damage.

One-way system

A linear economy takes resources and uses them, often creating large amounts of pollution and waste at all stages of a product's life cycle.

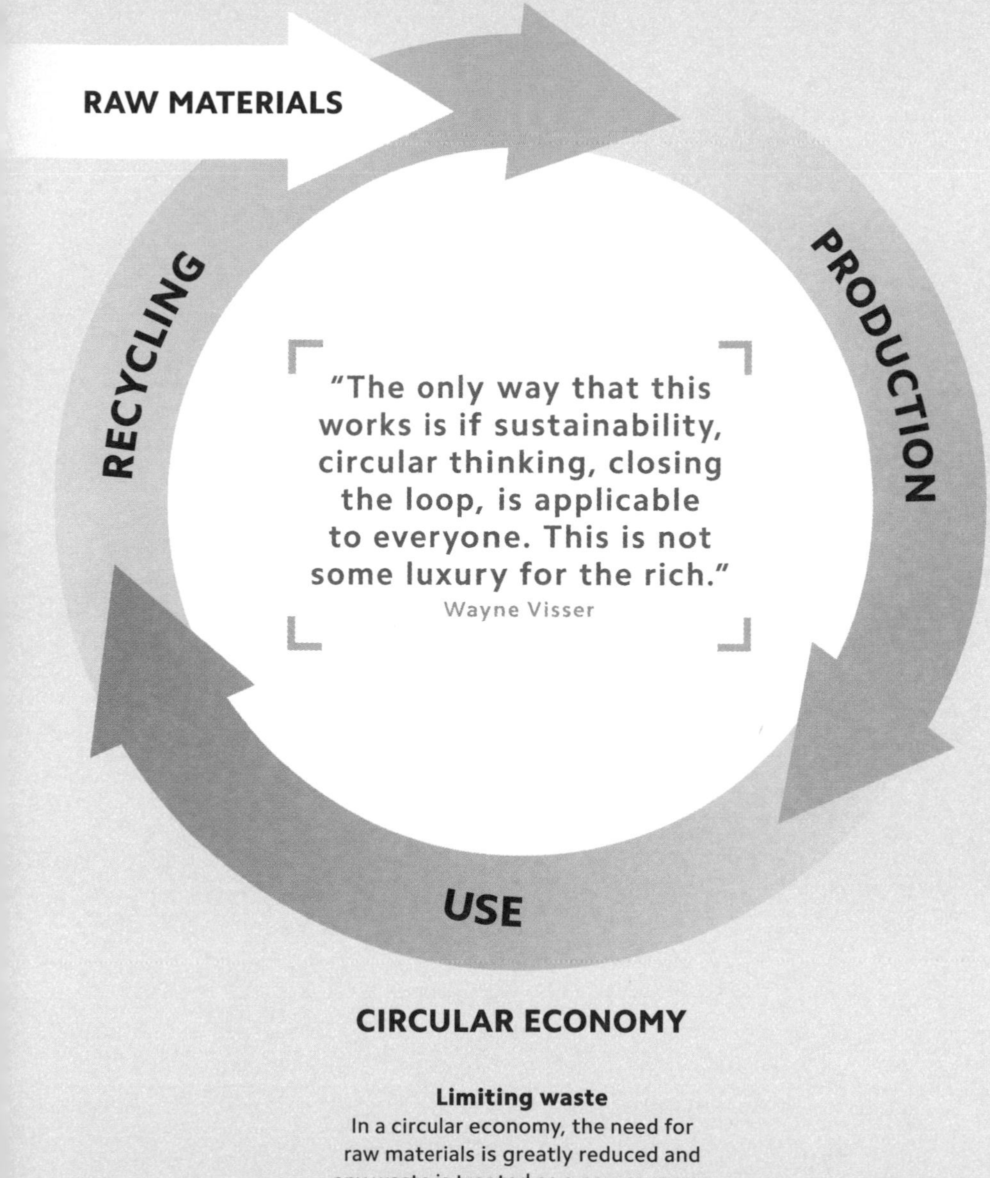

RAW MATERIALS

PRODUCTION

USE

RECYCLING

"The only way that this works is if sustainability, circular thinking, closing the loop, is applicable to everyone. This is not some luxury for the rich."

Wayne Visser

CIRCULAR ECONOMY

Limiting waste

In a circular economy, the need for raw materials is greatly reduced and any waste is treated as a new resource to be fed back into the economy.

PRECIOUS RESOURCE

Freshwater scarcity has driven the need to create better water management technology, such as small-scale desalination, and energy-efficient, nonpolluting wastewater treatment and purification.

SLASHING EMISSIONS

Transporting people and goods contributes greatly to carbon emissions, but the increasing use of electric-powered vehicles and biofuels is helping to reduce the reliance on fossil fuels.

GREEN PRODUCTS

Materials science is striving to create new products that involve less polluting production processes and stay out of the waste stream for longer by being more durable or, if disposed of, fully recyclable.

CLEAN TRANSPORTATION

CLEAN WATER

CLEAN MATERIALS

CLEANING UP OUR TECHNOLOGIES

Clean technology is a loosely defined sector involving products, processes, and services that reduce negative environmental impacts—from better wastewater treatment to new biofuels or recycling techniques. Growth in areas such as solar and wind power, energy efficiency, and green transportation, has seen the clean technology sector expand greatly. In addition, there are many ways in which clean technology is an intermediary step to making consumption less environmentally harmful. For example, dematerializing the production of drinks cans has led to less aluminum being used, and some digital products have replaced physical products.

CLEAN AND EFFICIENT

Great strides have already been made in developing renewable energy sources and in making existing technology more energy efficient and less polluting.

Tools to survive

Clean technology can make a big difference in fighting climate change, from making better use of resources to decreasing or eliminating waste and greenhouse emissions.

GREEN BY DESIGN

Energy consumption and greenhouse gas emissions can be reduced through greater efficiencies in energy use and the production of materials. Material efficiencies include reducing waste, smarter product design, and using recycled metals that require 60–90 percent less energy intensity than primary metal production from metal ores. Energy efficiencies include better insulation and ventilation to reduce HVAC (heating, ventilation, and air conditioning), smart lighting, and more fuel-efficient vehicles achieved through improved aerodynamics, engine design, and weight reduction.

"Energy efficiency is a major lever for reducing CO_2 emissions along all parts of the energy chain, from the production of resources all the way to final consumption."

Joe Kaeser

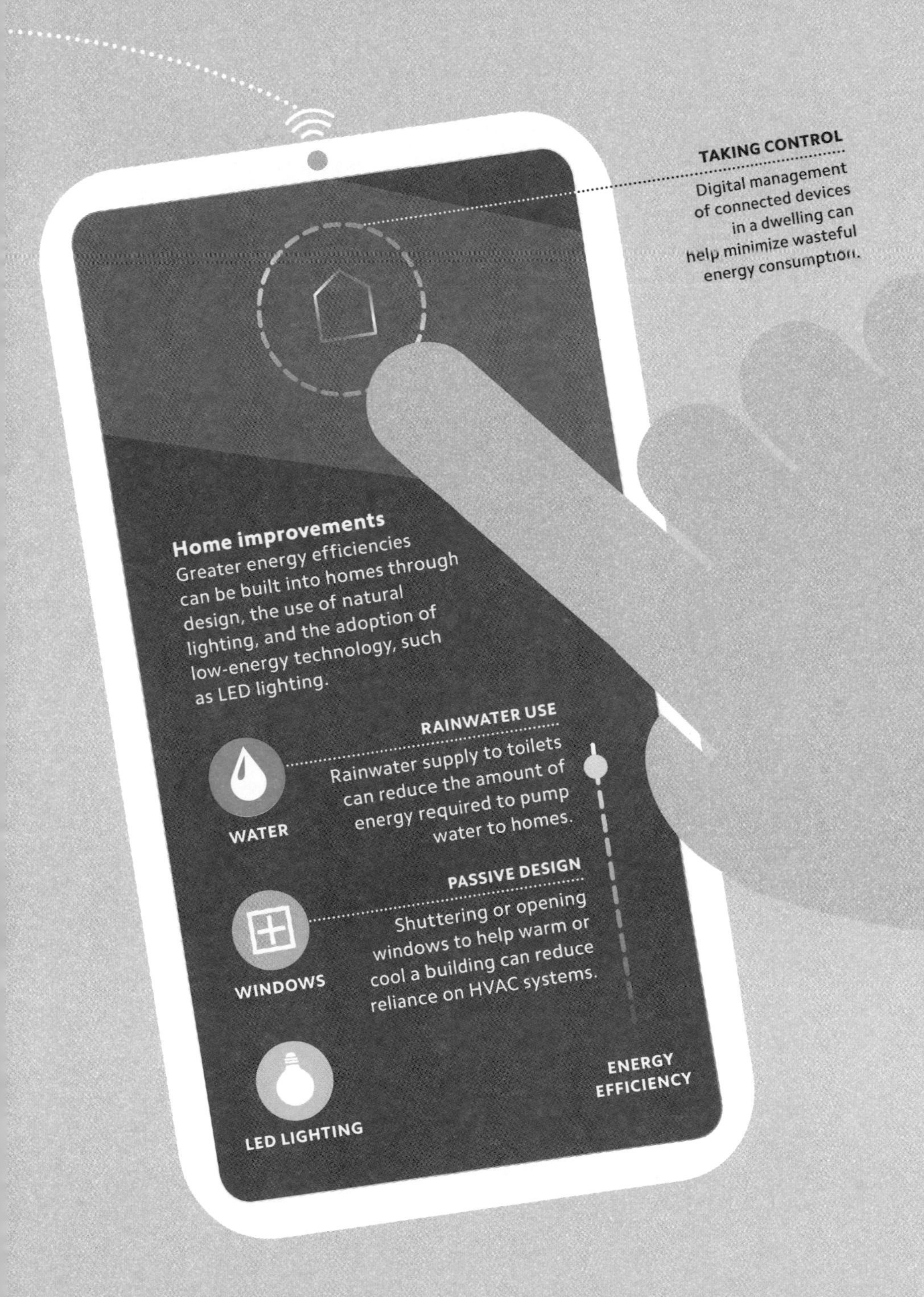
TAKING CONTROL
Digital management of connected devices in a dwelling can help minimize wasteful energy consumption.
Home improvements
Greater energy efficiencies can be built into homes through design, the use of natural lighting, and the adoption of low-energy technology, such as LED lighting.
RAINWATER USE
Rainwater supply to toilets can reduce the amount of energy required to pump water to homes.
WATER
PASSIVE DESIGN
Shuttering or opening windows to help warm or cool a building can reduce reliance on HVAC systems.
WINDOWS
LED LIGHTING
ENERGY EFFICIENCY

> "The answers to how to live sustainably on our planet are all around us."
> Janine Benyus

Generating change
Global energy consumption is not decreasing, but moving to cleaner and more efficient technology can help mitigate climate change.

A NEW WAVE OF INNOVATION

The use of renewable energy is soaring as technologies improve and societies begin to move away from fossil fuels. In fact, solar (see opposite) and wind (see p.132) electricity combined is projected to overtake coal and gas by 2024. Renewables present challenges for our electricity grids, due to their intermittent supply. However, they also allow for widespread production. This could bring electricity to people previously off-grid—effectively leapfrogging fossil fuels.

GREEN ENERGY FROM THE SUN

Photovoltaic systems produce electricity from the sun—the energy source for virtually all life. In some parts of the world, solar energy now provides the cheapest electricity in history. Panels can be installed everywhere, from windows to road surfaces, which means they can be integrated almost anywhere and do not require large-scale land use. Other technologies focus the sun's rays to produce heat. On small scales, this can cook food, but on large scales it can generate heat above 1,830°F (1,000°C).

Fueled by photons
When light particles (photons) hit a photovoltaic panel, the energy from the photons frees electrons in the panel, allowing them to flow through a circuit.

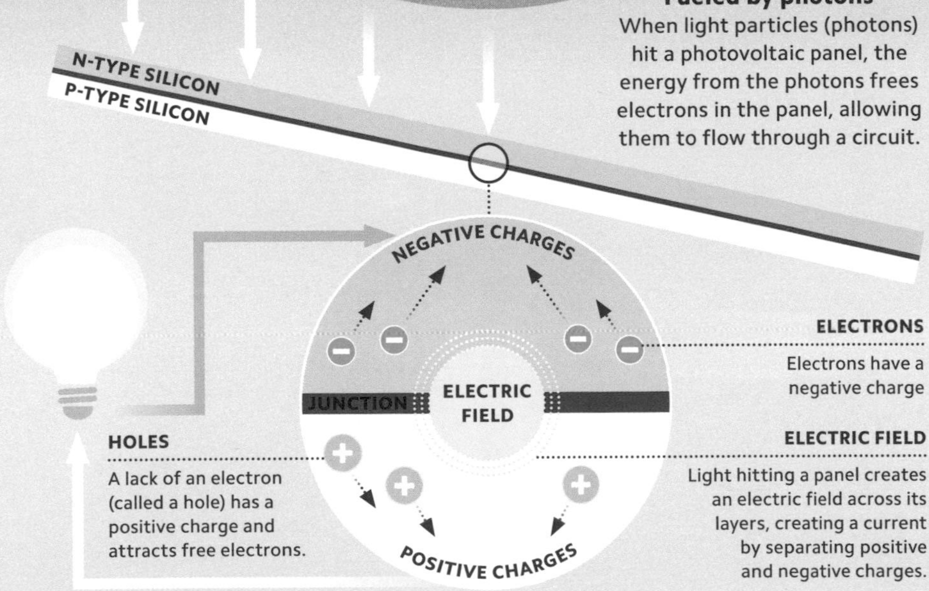

ELECTRONS
Electrons have a negative charge

HOLES
A lack of an electron (called a hole) has a positive charge and attracts free electrons.

ELECTRIC FIELD
Light hitting a panel creates an electric field across its layers, creating a current by separating positive and negative charges.

FISSION FOR ENERGY

Some 450 nuclear reactors in 30 countries produce one-tenth of the world's electricity. This involves atomic fission (splitting atoms' nuclei to release energy) of small quantities of enriched uranium fuel. The process provides reliable, low-carbon electricity with emissions, mostly from mining, processing, and transporting uranium. Public mistrust heightened by disasters, the cost of decommissioning old reactors, and the challenges of storing radioactive waste, has halted or reversed nuclear power's growth in many nations.

NOT CARBON NEUTRAL

Nuclear plants have a small climate impact from power generation itself, but are not considered to be carbon neutral due to the mining and processing of uranium.

RADIOACTIVE WASTE

Nuclear waste remains radioactive for thousands of years, making it hard to dispose of safely.

TOWER TROUBLE

It takes about 14–15 years to build a nuclear power plant, and the costs of building and decommissioning a plant remain very high.

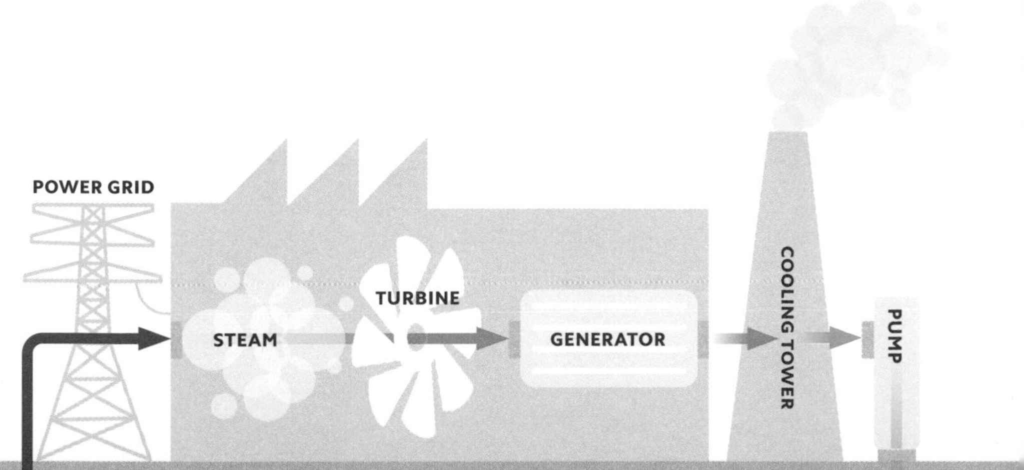

POWER GRID

TURBINE

STEAM

GENERATOR

COOLING TOWER

PUMP

GOING UNDERGROUND

Steam heat
Cold water is pumped deep below the surface where hot rocks heat the water and create steam. Water and steam return to the surface in pipes to drive turbines in the power plant.

Reliable and renewable, tapping into heat energy below ground to generate electricity produces as little as one-twentieth of the carbon emissions of a fossil fuel power plant. Iceland generates 27 percent of its electricity using geothermal energy—one of just a handful of nations including New Zealand, Kenya, and the Philippines where it is used to a great degree. Barriers to wider adoption include high start-up costs, limited site suitability, and concerns about geothermal operations increasing earthquake frequency.

HOT WATER IS PUMPED UP

COLD WATER IS PUMPED DOWN

HOT GRANITE

ENERGY FROM THE AIR

Humans have used windmills for more than a millennium, harnessing the power of prevailing winds to carry out tasks such as pumping water from wells or grinding flour. This technology is now producing electricity. Wind turbines can be built on land or offshore, where wind is faster and more constant. Some of today's turbines are truly vast—approaching the Eiffel Tower's height—and able to power thousands of homes.

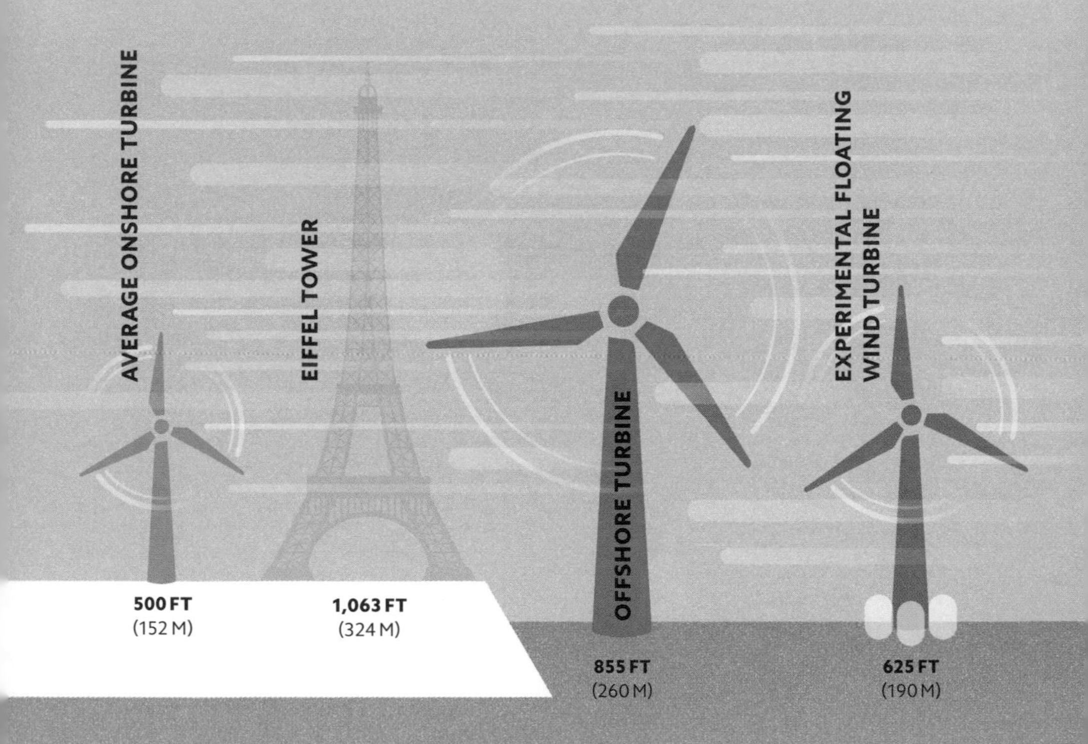

Growing capacity
Advances in technology mean that wind turbines are becoming larger and more efficient. Offshore capacity continues to grow, while onshore capacity growth has slowed as new sites become more difficult to find.

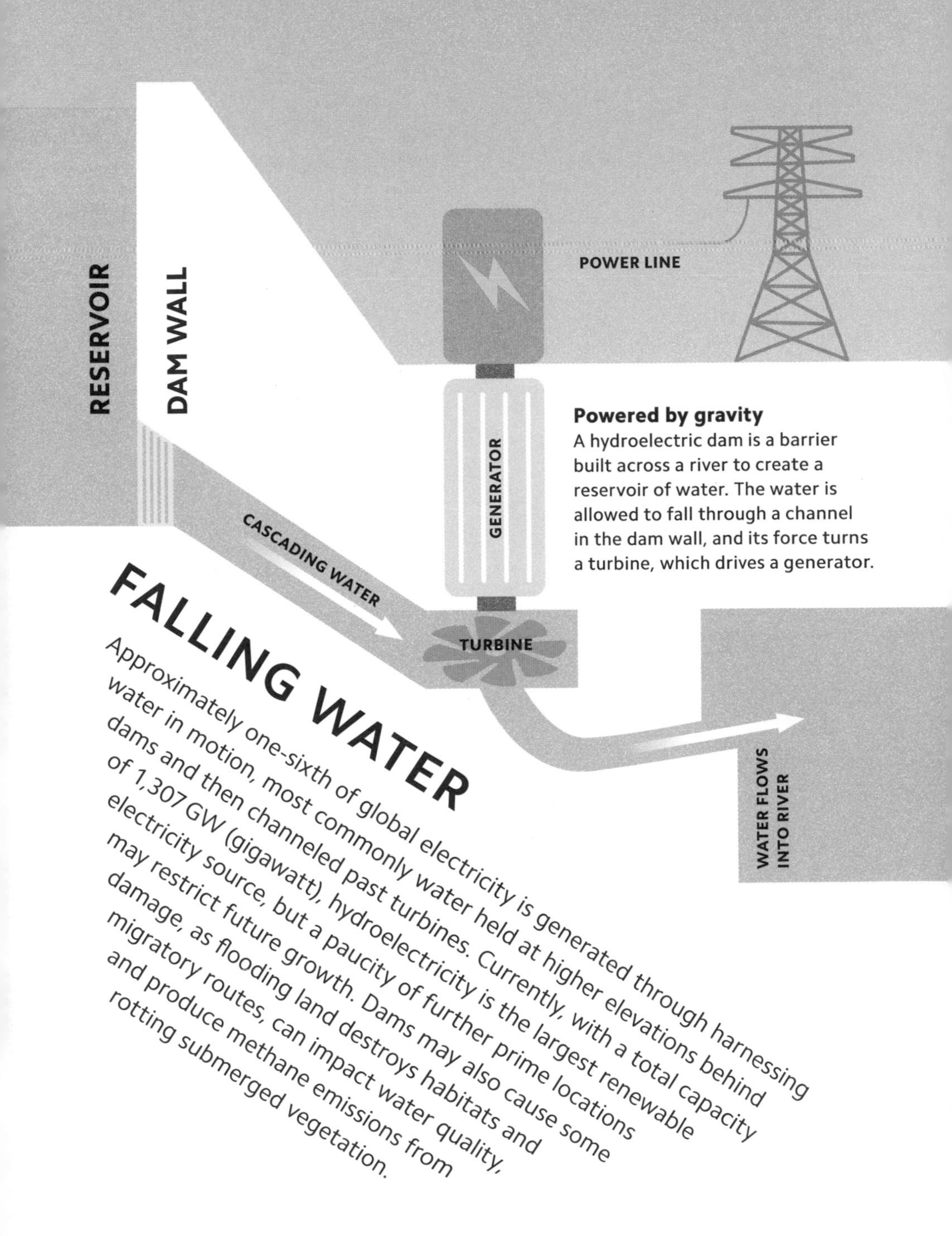

Powered by gravity
A hydroelectric dam is a barrier built across a river to create a reservoir of water. The water is allowed to fall through a channel in the dam wall, and its force turns a turbine, which drives a generator.

FALLING WATER

Approximately one-sixth of global electricity is generated through harnessing water in motion, most commonly water held at higher elevations behind dams and then channeled past turbines. Currently, with a total capacity of 1,307 GW (gigawatt), hydroelectricity is the largest renewable electricity source, but a paucity of further prime locations may restrict future growth. Dams may also cause some damage, as flooding land destroys habitats and migratory routes, can impact water quality, and produce methane emissions from rotting submerged vegetation.

SEA ENERGY

Created by winds whipping across ocean surfaces, waves have huge, largely untapped, potential. Various floating, submerged, or coastally-sited devices can transform waves' kinetic energy into electricity, but producing commercial-scale devices to operate in such a harsh, variable environment remains a challenge. The rewards, though, could be significant. An estimated 2.64 trillion kWh (kilowatt hours) of electricity—64 percent of US demand in 2019—is estimated to exist within the waves of US coastal waters, for example.

Attenuator
One type of device that generates electricity from undulating wave motion is called an attenuator. The attenuator relays power to shore via a cable running along the sea floor.

HINGED SECTIONS
CONSTANT MOTION
Attenuators can be placed in areas where the ocean is moving all the time, creating a steady source of kinetic energy.
HYDRAULIC ACTION
Floating sections are connected by hydraulic pipes. The wave motion pumps oil at high pressure through the motor, which drives the generator and creates electricity.
MOTOR
GENERATOR
WIRED TETHER

HIGHS AND LOWS

Carbon-free and renewable, but a potential threat to estuarine ecosystems, tidal barrage dams across basins and lagoons exploit the predictable flow of water in high and low tides. The water travels through sluices where it rotates the electricity-generating turbines. Adoption remains rare due to the high cost of construction, the tides' intermittence, and environmental impacts that include lowering salinity and creating physical barriers that restrict the free movement of marine species. Alternatives to barrages include individual turbines placed in fast-flowing tidal streams.

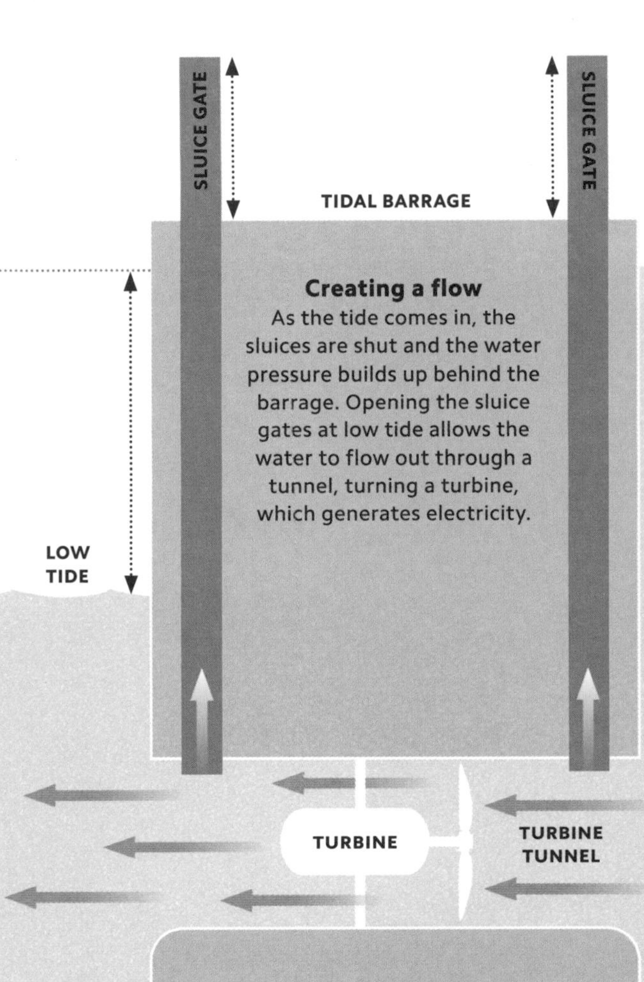
SLUICE GATE
TIDAL BARRAGE
SLUICE GATE
LOW TIDE
TURBINE
TURBINE TUNNEL
HIGH PRESSURE
Sluice gates hold water at the high-water level. which creates pressure on the barrage wall as the water on the other side recedes with the tide.
HIGH TIDE

Creating a flow
As the tide comes in, the sluices are shut and the water pressure builds up behind the barrage. Opening the sluice gates at low tide allows the water to flow out through a tunnel, turning a turbine, which generates electricity.

CARBON ABSORPTION (FIXATION)

GROWING TREES

The rate of net deforestation (see p.52) has slowed, in part due to concerted tree-growing drives: either restocking existing forests (reforestation) or creating new tree cover (afforestation). Among the multiple benefits—from combating soil erosion to providing employment—is expanding forests' vital role as carbon sinks. Trees lock away carbon by capturing atmospheric CO_2 and transforming it into biomass. Millions of hectares of new trees reaching maturity are required, though, for a major impact on atmospheric CO_2 levels.

O_2

Carbon sponge
Some 20–30 percent of a tree's biomass is underground in its anchoring and water- and nutrient-supplying root system. The root system stores much of the carbon taken by the tree from the atmosphere.

OXYGEN RELEASE
During photosynthesis, plants combine carbon dioxide and water to make glucose. The waste product of this process is oxygen.

PART OF THE CARBON CYCLE
Roots release compounds containing carbon into the soil and the decomposition of dead roots also releases carbon. Some carbon is then released from the soil into the atmosphere.

NURTURING NATURE

Many ecosystems can recover after being disturbed and damaged by natural forces, such as flooding, or human impact, such as logging, overgrazing, or pollution. Methods to help kickstart or accelerate recovery can involve removal of the cause of the disturbance, cleanup and revegetation, and reintroduction of species that have been lost. In some cases, conditions have changed too much for rollback and a different ecosystem is created and managed.

Giving nature a helping hand
Removing pollution sources, enriching depleted soils, and replanting key native tree and plant species can encourage a formerly degraded ecosystem to recover.

FRESH START
A carefully restored landscape may, over time, attract an array of species necessary to produce a vibrant ecosystem.

RECOVERING HABITAT

PLANTS

INSECTS

BIRDS

TREES

ANIMALS

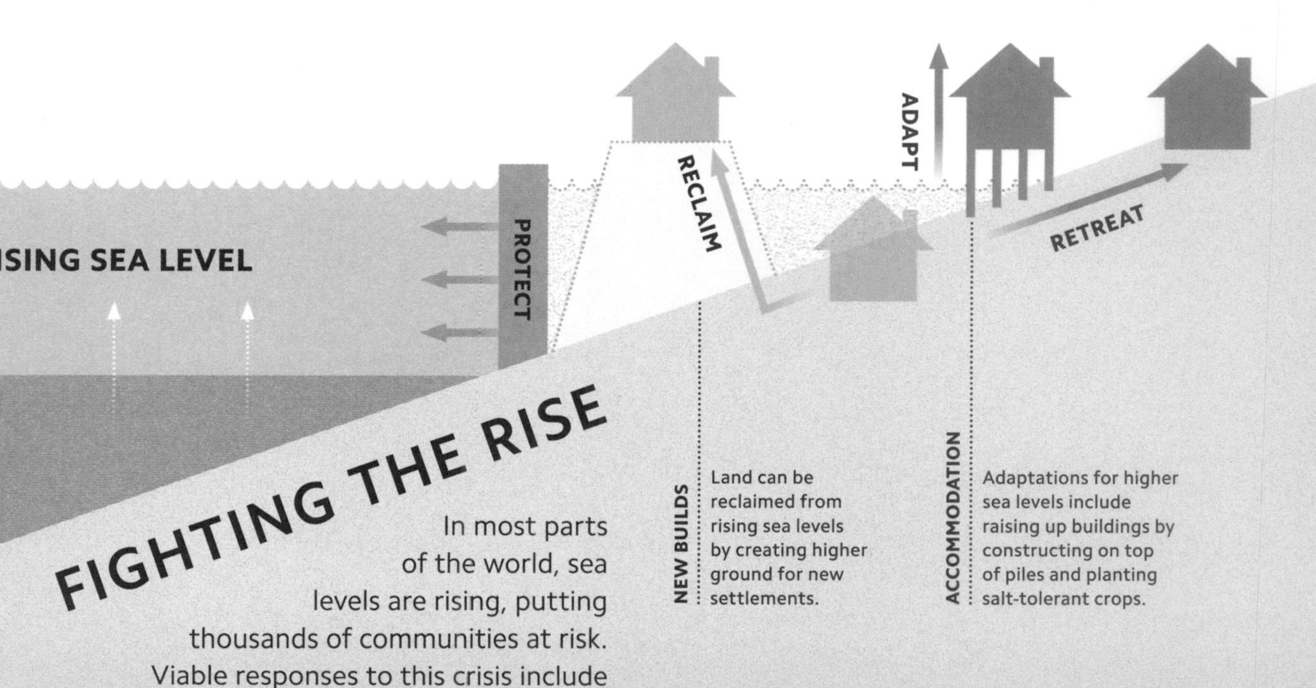

RISING SEA LEVEL
PROTECT
RECLAIM
ADAPT
RETREAT
NEW BUILDS
Land can be reclaimed from rising sea levels by creating higher ground for new settlements.
ACCOMMODATION
Adaptations for higher sea levels include raising up buildings by constructing on top of piles and planting salt-tolerant crops.

FIGHTING THE RISE

In most parts of the world, sea levels are rising, putting thousands of communities at risk. Viable responses to this crisis include regulatory options, such as land use change, and structural options, such as the construction of hard structures (concrete sea walls, or dykes) or soft structures (levees or dunes). Significant rises in sea level may see the creation of no-build zones in high-risk areas, or even the abandonment of coastal areas along with a retreat to higher land.

Varied solutions

The potential solutions to rising sea levels are location-specific, and each solution must be measured against its cost and long-term effectiveness. Building sea walls, for example, is quick, yet costly, and may not offer long-term protection.

BEFORE THE STORM

Tropical storms (see p.76) are becoming more intense due to climate change. Early-warning systems can give advance information of a storm's likely path and its predicted zone of landfall. Armed with this information, response teams working alongside law enforcement, or even armed services, erect temporary shelters, broadcast warnings, or advise mass evacuations for those living in particularly vulnerable areas. More commonly, however, people are instructed to remain indoors and wait for the storm to pass.

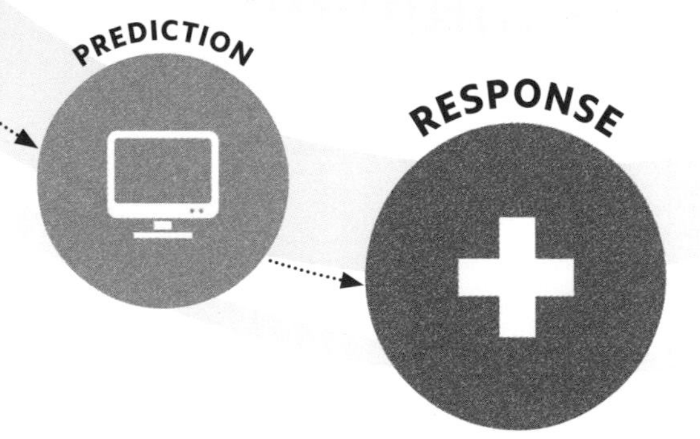

Storm warning
Reconnaissance tools, such as radar and satellites, monitor developing storms. The data is then run through computer models to determine the best emergency response measures.

FIGHTING THE FLAMES

Climate change and the rise in temperature can be linked to an increase in the number and intensity of wildfires around the world (see p.84). Wildfires often spread rapidly and can be hard to control or contain. Key strategies to prevent the impacts of wildfires include creating and maintaining firebreaks, managing areas where there is a natural buildup of dry vegetation that can be fuel for a fire, and establishing an early warning system to spot signs of fire and alert firefighters.

FIRE SAFETY

Wildfires may have natural causes, such as lightning strikes, but some are caused by human activity. Public awareness of fire safety rules is essential.

> **"Climate is really running the show in terms of what burns."**
> Park Williams

CONTAINING FIRES

A firebreak is an area that has been cleared of fuel and combustible vegetation, or where there is a natural break in vegetation. Firebreaks help to stop the rapid spread of wildfires.

AERIAL SUPPORT

Thousands of gallons of water or fire retardant can be dropped onto a blazing forest fire to bring the fire under control.

Putting out a fire

Firefighting techniques focus on depriving a fire of at least one element of the fire triangle—fuel, heat, and oxygen.

CHANG
A PERS
SCALE

E ON
ONAL

Individual, everyday activities add up to global-scale impacts. In richer countries, the average carbon footprint is often many times greater than in poorer parts of the world. Whether it's how we get around or what we eat, making greener choices can substantially lower our carbon footprints. Along with various forms of climate activism, personal shifts can change the climate conversation. That can encourage others to make a difference, while putting pressure on both governments and companies to implement the large-scale shifts (see pp.114–141) that are also needed.

CHANGE FROM THE GROUND UP

Popular in 1970s environmentalism, the phrase "Think Globally, Act Locally" has recently been readopted to signify the importance of implementing climate change-tackling strategies from the ground up at local community levels. This is viewed as a way to initiate meaningful change and apply pressure upward. Cities are considered especially important in this approach, with many initiating schemes or setting their own low-carbon targets. The largest global alliance for climate change, the Global Covenant of Mayors (GCoM), has seen more than 10,000 cities and local governments commit to reduce emissions by 2030.

CITY INITIATIVES

APPLYING PRESSURE

Individuals banding together in their local communities can pressurize city administrators to take meaningful action on climate change.

COMMUNITY ACTION

FORMING A GLOBAL NETWORK

The cooperation of city initiatives is required in order to fulfill global goals.

GLOBAL COVENANT

International collaboration
Collective action can have real global results; GCoM groups representing 800 million people in 138 countries have promised a reduction in emissions.

"Although the magnitude of climate change may make individuals feel helpless, individual action is critical for meaningful change."
Mia Armstrong

MAKING THEIR VOICES HEARD

Propelled by increasingly stark predictions, a perceived lack of progress, and a surge in engagement among the young (see p.149), climate activism has recently grown in scale and intensity. Some activists have taken high-profile direct action, including the occupation of buildings, bridges, and oil platforms. Most have engaged in strikes, protests, and demonstrations, often coordinated and publicized through social media. During the global climate strikes of 2019, for example, more than 6,000 organized protests and events took place.

Calls for change
Climate protests give a voice to concerned citizens united by their vision of a better world. These demonstrations can prompt conversations, promote understanding, and help apply political pressure to institutions and governments.

CIVIL DISOBEDIENCE

GRASSROOTS MOVEMENTS

INDIVIDUAL ACTION

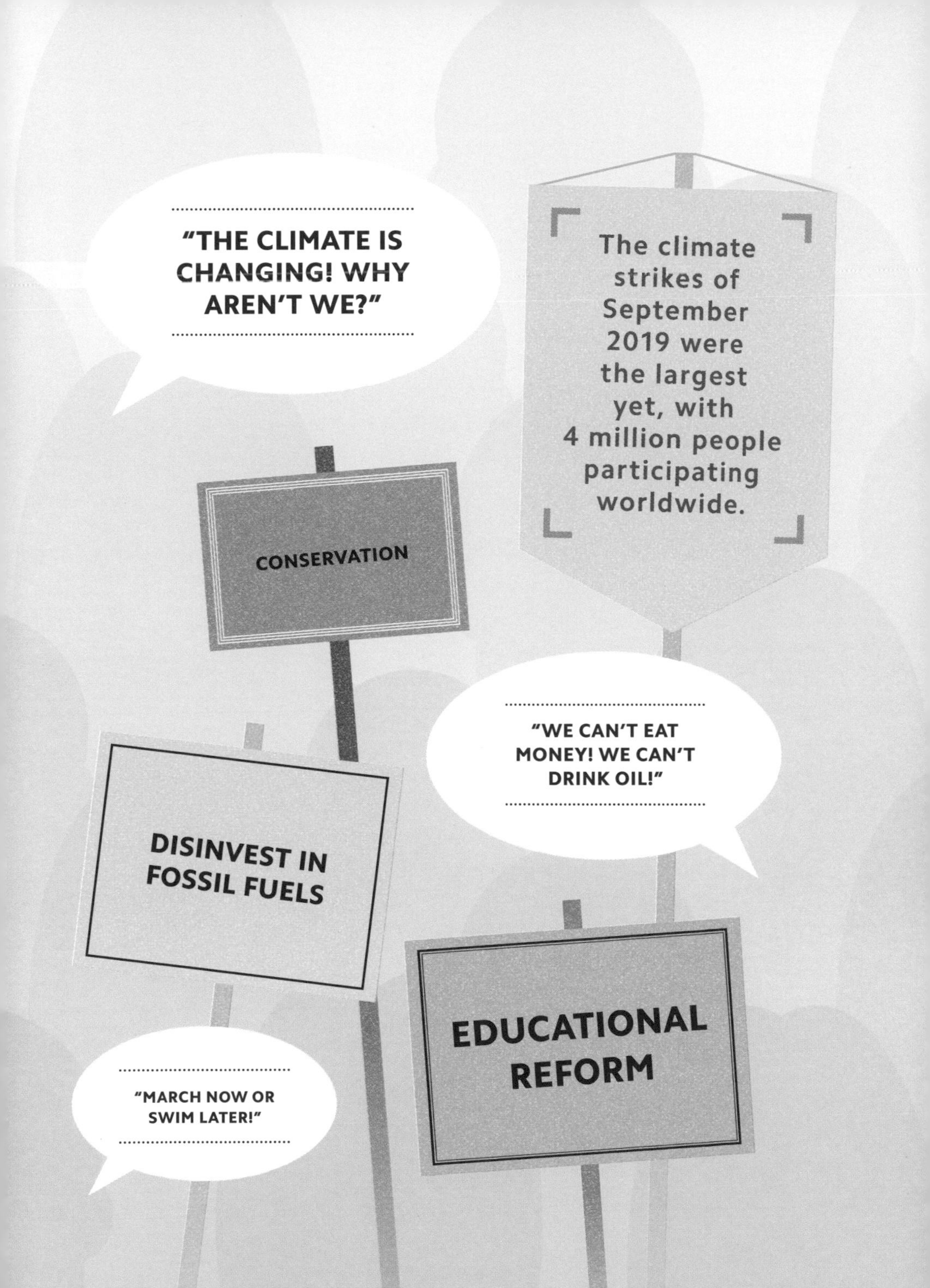

"THE CLIMATE IS CHANGING! WHY AREN'T WE?"

The climate strikes of September 2019 were the largest yet, with 4 million people participating worldwide.

CONSERVATION

"WE CAN'T EAT MONEY! WE CAN'T DRINK OIL!"

DISINVEST IN FOSSIL FUELS

EDUCATIONAL REFORM

"MARCH NOW OR SWIM LATER!"

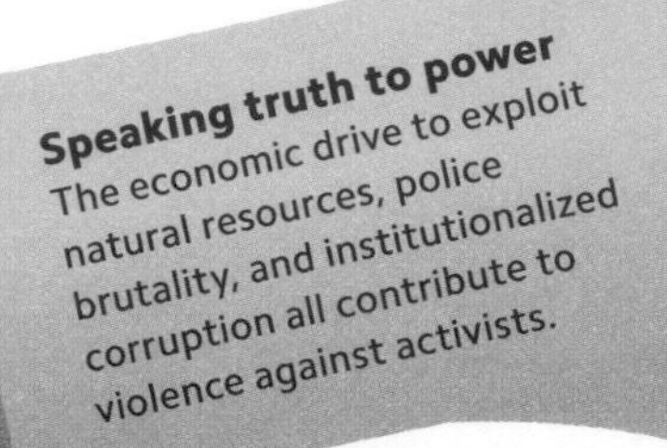

Speaking truth to power The economic drive to exploit natural resources, police brutality, and institutionalized corruption all contribute to violence against activists.

"I can't stay silent faced with all that is happening to my people."
Jakeline Romero Epiayu

FIGHTING FOR THEIR LIVES

The most dangerous type of climate activism is usually on the frontline, where participants strive to protect the rights and well-being of their local communities. Frontline activist issues include illegal deforestation, land grabs, and pollution. Increasingly, the challenge that some activists pose to the interests of big business and government can result in them facing imprisonment, intimidation, violence, or even death.

STOP BURNING OUR FUTURE

In 2018, following Sweden's hottest summer in 262 years, 15-year-old Swedish activist Greta Thunberg began a solo school strike. Her pledge to strike once a week until Sweden lowered carbon emissions in line with the Paris Agreement sparked a global movement. Huge school strikes and protests the following year revealed a passionate network of students and children around the world committed to addressing climate change. In 2020, however, the campaigning went digital due to coronavirus pandemic restrictions.

Teaching the world
Young climate change activists often highlight that they are the ones who will inherit a damaged planet. They emphasize climate justice, science-based approaches, and the limiting of emissions.

CLIMATE JUSTICE FOR ALL

FOLLOW THE PARIS AGREEMENT

UNITE BEHIND THE SCIENCE

SAFE PATHWAY UNDER 1.5°C

ALL IN THIS TOGETHER

In the fight against the climate crisis, a collective, worldwide effort is needed. As climate change's most devastating impacts do not recognize national borders, so scientists, politicians, companies, and change-makers need to reach across national and political divides. Greater partnerships, agreement, and collaboration must be forged between all parties, from individuals to governments, from lower-income to higher-income economies, and from business groups to environmental groups.

YOUR VOTE

For those living in a democracy, casting a vote for representatives with environmentally friendly policies can make a real difference.

YOUR VOICE

If climate change is to be addressed, then speaking up and raising awareness is a fundamental part of meaningful action.

"Out of this crisis can come a collective reconsideration of our priorities. How to live sustainably on a finite planet with finite space, food, and water."

Michael E. Mann

How to take action

Collective change is only possible when individuals, en masse, use the tools at their disposal to raise awareness of the climate crisis and combat it on a daily basis.

YOUR TIME

Very few people are able to fight climate change all day, every day. Most people must simply assign as much time as they can to taking action.

YOUR MONEY

Your money, what you buy and where you spend it, often has a great impact on what your society looks like and what it values.

ACTIVE TRAVEL

Replacing some car-based journeys with active ways of traveling, such as walking and cycling, is good for the environment and also has physical health benefits.

Small change, big difference

Studies have found that even small changes, such as replacing one car trip per day with cycling, can have a significant cumulative effect on reducing a person's carbon footprint. Cycling is also more cost-effective than driving.

CHANGING MINDSETS

The private car has come to dominate transportation, while flying has become a normality for many. Putting the brakes on climate change means rethinking and reshaping these relationships. Individuals can take actions, such as walking or cycling and avoiding flying. However, infrastructure must also change to support both new technology and public transportation (see p.155). Many changes are already underway, as electric cars (see p.154) are bought in record numbers, and many countries begin phasing out conventional vehicles.

REDUCE WASTE

Some easy steps can be taken to reduce waste associated with food, such as buying and consuming only what you need, freezing fresh food to eat at a later date, and buying food with minimal packaging.

EATING GREEN

GROW YOUR OWN

REDUCE FOOD WASTE

SOURCE RESPONSIBLY

EAT A MORE DIVERSE DIET

CHOOSE SUSTAINABLE

EAT FOODS IN SEASON

REDUCE MEAT CONSUMPTION

REDUCE PACKAGING

Making our dinner plates greener can improve the health of both people and planet. Swapping red meat for plant-based proteins, such as beans, nuts, and tofu, can shrink emissions a hundredfold. And plant milks (such as oat and nut) can be three times greener than dairy. Buying local might seem crucial, but transport only represents a small part of our food's footprint. So, what we eat matters more than where it's from.

EAT LESS MEAT

Meat and dairy farming uses a lot of land and water. Livestock farming also generates greenhouse gas emissions, so reducing the demand for meat and dairy will have a big impact on the climate.

RESPONSIBLE DIVERSITY

Nearly 75 percent of the world's food supply (see pp.48–49) comes from 12 plants and five animal species. This causes a threat not just to the environment, but also to food security (see p.112).

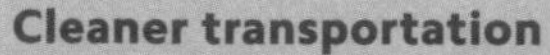

Cleaner transportation
Sophisticated control systems allow some cars to convert movement energy back into stored electric energy, such as when braking. New innovation also means that cars will soon be able to feed electricity back into power grids.

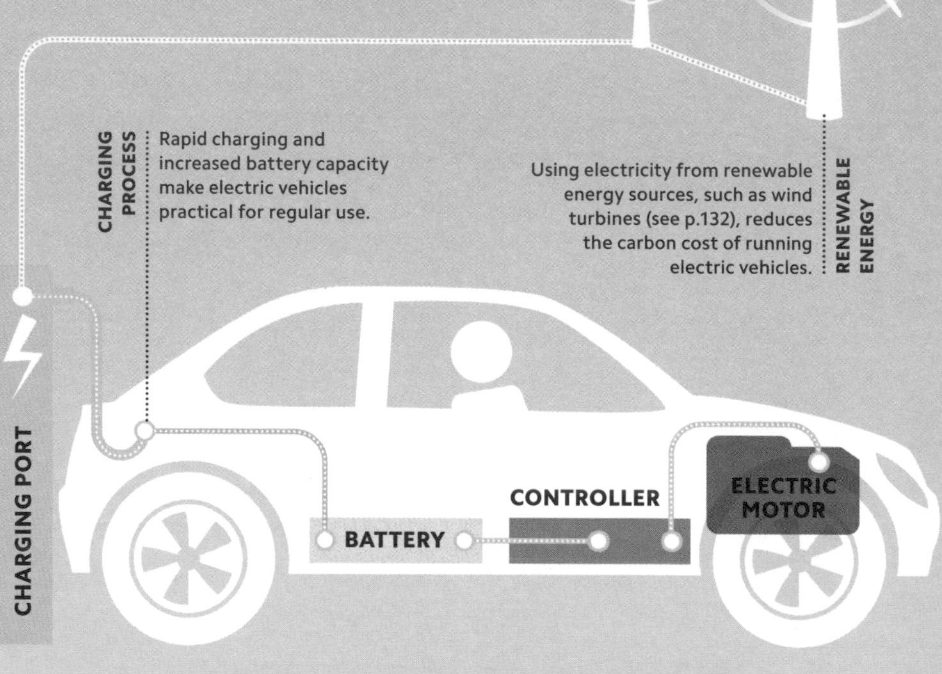

CHARGING PROCESS
Rapid charging and increased battery capacity make electric vehicles practical for regular use.
RENEWABLE ENERGY
Using electricity from renewable energy sources, such as wind turbines (see p.132), reduces the carbon cost of running electric vehicles.
CHARGING PORT
BATTERY
CONTROLLER
ELECTRIC MOTOR

DRIVING ELECTRIC

Compared to the lifetime pollution from conventional, internal-combustion engine cars, electric vehicles can produce hundreds of times less waste and a third of the greenhouse gases. What's more, as electricity grids become greener, so do the cars. Battery technology is improving every year, allowing cars to cover longer ranges per charge and reducing charging times. These cars do not have a tailpipe, which means that they do not contribute to the air pollution in our cities.

TRAVELING TOGETHER
1 BUS CAN REMOVE
30 CARS FROM THE ROAD

Not all clean changes to transportation need to be high-tech. Public transportation, such as a bus or local train, can produce just a sixth of the emissions per passenger of a similar journey made by a car. And the benefits can be even more pronounced for longer-distance travel on buses and trains. Combining these shared travel options with new technology—for example, using electric buses—can further slash emissions, while providing cleaner, quieter streets.

Taking the bus

Public transportation systems not only ease urban congestion and cut emissions, studies show that they can have a positive "green" economic impact for cities and help to create a better quality of life.

Every $1 invested in public transportation could generate $5 in economic returns.

INDEX

Page numbers in **bold** refer to main entries.

D

E

F

T

UVWY

ACKNOWLEDGMENTS

DK would like to thank the following for their help with this book: Joy Evatt for proofreading; Helen Peters for the index; Senior Jacket Designer Suhita Dharamjit; Senior DTP Designer Harish Aggarwal; Jackets Editorial Coordinator Priyanka Sharma; Managing Jackets Editor Saloni Singh.

Reference sources
18–19: United States Geological Survey; **31:** Scripps CO2 program (https://scrippsco2.ucsd.edu/data/atmospheric_co2/primary_mlo_co2_record.html); **42:** UN Department of Economic and Social Affairs, Population Division (2017). World Population Prospects 2017 – Data Booklet; **66** Ellen MacArthur Foundation A NEW TEXTILES ECONOMY; **88** NASA; **89** NASA, National Snow and Ice Data Centre; **96**: NASA; **121** J Lokrantz-Azote based on Steffen et al 2015; **155** European Automobile Manufacturers' Association (ACEA) https://www.acea.be/automobile-industry/buses).

All images © Dorling Kindersley
For further information see: www.dkimages.com